Methods in Molecular Biology

Series Editor
John M. Walker
School of Life and Medical Sciences
University of Hertfordshire
Hatfield, Hertfordshire, UK

For further volumes:
http://www.springer.com/series/7651

For over 35 years, biological scientists have come to rely on the research protocols and methodologies in the critically acclaimed *Methods in Molecular Biology* series. The series was the first to introduce the step-by-step protocols approach that has become the standard in all biomedical protocol publishing. Each protocol is provided in readily-reproducible step-by-step fashion, opening with an introductory overview, a list of the materials and reagents needed to complete the experiment, and followed by a detailed procedure that is supported with a helpful notes section offering tips and tricks of the trade as well as troubleshooting advice. These hallmark features were introduced by series editor Dr. John Walker and constitute the key ingredient in each and every volume of the *Methods in Molecular Biology* series. Tested and trusted, comprehensive and reliable, all protocols from the series are indexed in PubMed.

Abscisic Acid

Methods and Protocols

Edited by

Takuya Yoshida

Max-Planck-Institut für Molekulare Pflanzenphysiologie, Potsdam-Golm, Germany; Centre of Plant Systems Biology and Biotechnology, Plovdiv, Bulgaria; Lehrstuhl für Botanik, Technische Universität München, Freising-Weihenstephan, Germany

Humana Press

Editor
Takuya Yoshida
Max-Planck-Institut für Molekulare
Pflanzenphysiologie
Potsdam-Golm, Germany

Centre of Plant Systems Biology
and Biotechnology
Plovdiv, Bulgaria

Lehrstuhl für Botanik
Technische Universität München
Freising-Weihenstephan, Germany

ISSN 1064-3745 ISSN 1940-6029 (electronic)
Methods in Molecular Biology
ISBN 978-1-0716-2155-4 ISBN 978-1-0716-2156-1 (eBook)
https://doi.org/10.1007/978-1-0716-2156-1

This Humana imprint is published by the registered company Springer Science+Business Media, LLC, part of Springer Nature.
The registered company address is: 1 New York Plaza, New York, NY 10004, U.S.A.

Preface

The water status of plants is well managed at the organ, tissue, and cellular levels. The phytohormone abscisic acid (ABA) accumulates in response to cellular dehydration and serves as the key signaling molecule for water management in plants. In particular, ABA is involved in the regulation of seed maturation and acquiring desiccation tolerance, seed germination, stomatal closure, and stress-responsive gene expression. Since the identification of cytosolic ABA receptors and coreceptors, their target kinases, and ABA transporters in the late 2000s and the early 2010s, our knowledge about the molecular cascade of the ABA signaling pathway has been expanding. Moreover, the core components of ABA signaling have recently been suggested to be controlled by higher-order mechanisms, such as post-translational modifications and long-distance signals. ABA is essential in vascular plants as well as other phytohormones, and the roles of ABA have been widely studied in model species (i.e., *Arabidopsis thaliana*), crops (such as rice and maize), and trees.

This volume is intended to present a comprehensive overview of molecular biological methods to study ABA signaling, transport, and metabolism in several plant species: biochemical and molecular biological characterization of ABA receptors, coreceptors, and their target kinases (Chapters 1–3), characterization of post-translational modifications of the core signaling components (Chapters 4 and 5), assay systems for ABA transporters (Chapters 6 and 7), techniques to address long-distance signaling (Chapters 8 and 9), quantification and visualization of ABA levels (Chapters 10–12), and comprehensive *omics* approaches (Chapters 13–16). ABA signaling, transport, and metabolism are important potential targets for conventional and molecular breeding to develop crops that need less water during the vegetative and reproductive stages. To exemplify how basic and applied studies can be bridged, an approach to assess the effects of ABA agonists targeting ABA receptors in crops is also presented (Chapter 17).

Potsdam-Golm, Germany *Takuya Yoshida*

Acknowledgments

TY is supported by the European Union's Horizon 2020 research and innovation program, project PlantaSYST (SGA-CSA No. 739582 under FPA No. 664620).

Contents

Contributors

ABDULWAHED F. ALREFAEI • *Zoology Department, College of Science, King Saud University, Riyadh, Saudi Arabia*

BORJA BELDA-PALAZÓN • *Instituto Gulbenkian de Ciência, Oeiras, Portugal; GREEN-IT Bioresources for Sustainability, ITQB NOVA, Oeiras, Portugal*

MO-XIAN CHEN • *Co-Innovation Center for Sustainable Forestry in Southern China, College of Biology and the Environment, Nanjing Forestry University, Nanjing, Jiangsu Province, China; CAS Key Laboratory of Quantitative Engineering Biology, Shenzhen Institute of Synthetic Biology, Shenzhen Institutes of Advanced Technology, Chinese Academy of Sciences, Shenzhen, China*

ALBERTO COEGO • *Instituto de Biología Molecular y Celular de Plantas, Consejo Superior de Investigaciones Cientificas-Universidad Politecnica de Valencia, Valencia, Spain*

SEAN R. CUTLER • *Institute for Integrative Genome Biology, University of California, Riverside, CA, USA; Botany and Plant Sciences, University of California, Riverside, CA, USA*

GRAZYNA DOBROWOLSKA • *Institute of Biochemistry and Biophysics, Polish Academy of Sciences, Warsaw, Poland*

SANNA EHONEN • *Organismal and Evolutionary Biology Research Programme, Faculty of Biological and Environmental Sciences, University of Helsinki, Helsinki, Finland; Viikki Plant Science Centre, University of Helsinki, Helsinki, Finland*

BEI GAO • *Co-Innovation Center for Sustainable Forestry in Southern China, College of Biology and the Environment, Nanjing Forestry University, Nanjing, Jiangsu Province, China; State Key Laboratory of Desert and Oasis Ecology, Xinjiang Institute of Ecology and Geography, Chinese Academy of Sciences, Urumqi, China*

SUHYEON HEO • *Department of Biological Sciences, Sungkyunkwan University, Suwon, South Korea*

JOSE JULIAN • *Instituto de Biología Molecular y Celular de Plantas, Consejo Superior de Investigaciones Cientificas-Universidad Politecnica de Valencia, Valencia, Spain; Max Perutz Labs, Department of Biochemistry and Cell Biology, University of Vienna, Vienna, Austria*

JOOHYUN KANG • *Division of Integrative Biosciences and Biotechnology (IBB), POSTECH (Pohang University of Science and Technology), Pohang, South Korea*

YURI KANNO • *RIKEN Center for Sustainable Resource Science, Yokohama, Kanagawa, Japan*

BEOM-GI KIM • *Division of Metabolic Engineering, National Institute of Agricultural Sciences, RDA, Jeonju, Republic of Korea*

RIGYEONG KIM • *Division of Metabolic Engineering, National Institute of Agricultural Sciences, RDA, Jeonju, Republic of Korea*

EWA KRZYWINSKA • *Institute of Biochemistry and Biophysics, Polish Academy of Sciences, Warsaw, Poland*

SANGHO LEE • *Department of Biological Sciences, Sungkyunkwan University, Suwon, South Korea*

YEONGMOK LEE • *Department of Biological Sciences, Sungkyunkwan University, Suwon, South Korea*

Shengxue Liu • *College of Biological Sciences, China Agricultural University, Beijing, China*
Ying-Gao Liu • *State Key Laboratory of Crop Biology, College of Life Science, Shandong Agricultural University, Taian, Shandong, China*
Chong-Chong Lu • *State Key Laboratory of Crop Biology, College of Life Science, Shandong Agricultural University, Taian, Shandong, China*
Myung Ki Min • *Division of Metabolic Engineering, National Institute of Agricultural Sciences, RDA, Jeonju, Republic of Korea*
Feng Qin • *College of Biological Sciences, China Agricultural University, Beijing, China*
Pedro L. Rodriguez • *Instituto de Biología Molecular y Celular de Plantas, Consejo Superior de Investigaciones Cientificas-Universidad Politecnica de Valencia, Valencia, Spain*
Mohamed A. Salem • *Department of Pharmacognosy, Faculty of Pharmacy, Menoufia University, Menoufia, Egypt*
Mitsunori Seo • *RIKEN Center for Sustainable Resource Science, Yokohama, Kanagawa, Japan*
Kazuo Shinozaki • *Gene Discovery Research Group, RIKEN Center for Sustainable Resource Science, Ibaraki, Japan*
Maija Sierla • *Organismal and Evolutionary Biology Research Programme, Faculty of Biological and Environmental Sciences, University of Helsinki, Helsinki, Finland; Viikki Plant Science Centre, University of Helsinki, Helsinki, Finland*
Fumiyuki Soma • *Laboratory of Plant Molecular Physiology, Graduate School of Agricultural and Life Sciences, The University of Tokyo, Tokyo, Japan*
Katarzyna Patrycja Szymanska • *Institute of Biochemistry and Biophysics, Polish Academy of Sciences, Warsaw, Poland; Chair of Drug and Cosmetics Biotechnology, Faculty of Chemistry, Warsaw University of Technology, Warsaw, Poland*
Fuminori Takahashi • *Department of Biological Science and Technology, Faculty of Advanced Engineering, Tokyo University of Science, Tokyo, Japan; Gene Discovery Research Group, RIKEN Center for Sustainable Resource Science, Ibaraki, Japan; Graduate School of Industrial Science and Technology, Tokyo University of Science, Tokyo, Japan*
Taishi Umezawa • *Graduate School of Bio-Applications and Systems Engineering, Tokyo University of Agriculture and Technology, Tokyo, Japan; Faculty of Agriculture, Tokyo University of Agriculture and Technology, Tokyo, Japan*
Aditya S. Vaidya • *Institute for Integrative Genome Biology, University of California, Riverside, CA, USA; Botany and Plant Sciences, University of California, Riverside, CA, USA*
Rainer Waadt • *Institute of Technology, University of Tartu, Tartu, Estonia; Molekulargenetik und Zellbiologie der Pflanzen, Institut für Biologie und Biotechnologie der Pflanzen, Universität Münster, Münster, Germany*
Shunsuke Watanabe • *RIKEN Center for Sustainable Resource Science, Yokohama, Kanagawa, Japan*
Kazuko Yamaguchi-Shinozaki • *Laboratory of Plant Molecular Physiology, Graduate School of Agricultural and Life Sciences, The University of Tokyo, Tokyo, Japan; Research Institute for Agricultural and Life Sciences, Tokyo University of Agriculture, Tokyo, Japan*
Kota Yamashita • *Graduate School of Bio-Applications and Systems Engineering, Tokyo University of Agriculture and Technology, Tokyo, Japan*

AHMED ZAYED • *Department of Pharmacognosy, College of Pharmacy, Tanta University, Tanta, Egypt; Institute of Bioprocess Engineering, Technical University of Kaiserslautern, Kaiserslautern, Germany*

JIANHUA ZHANG • *Department of Biology, Hong Kong Baptist University, and State Key Laboratory of Agrobiotechnology, The Chinese University of Hong Kong, Shatin, Hong Kong, China*

FU-YUAN ZHU • *Co-Innovation Center for Sustainable Forestry in Southern China, College of Biology and the Environment, Nanjing Forestry University, Nanjing, Jiangsu Province, China*

Chapter 1

Inhibition of Type 2C Protein Phosphatases by ABA Receptors in Abscisic Acid–Mediated Plant Stress Responses

Yeongmok Lee, Suhyeon Heo, and Sangho Lee

Abstract

Abscisic acid (ABA) signaling pathway constitutes a key stress response mechanism in plants. Alternate interaction of a type 2C protein phosphatase (PP2C) with either an ABA receptor or a SNF1-related protein kinase 2 (SnRK2) is at the core of regulation of the ABA signaling pathway. Binding of an ABA receptor to a PP2C inhibits the enzymatic activity of the PP2C. Presence of multiple paralogs of ABA receptors and PP2Cs necessitates development of an inhibition assay. Here, we describe how to prepare the recombinant proteins of ABA receptors and PP2Cs for the inhibition assay and how to perform a colorimetry-based inhibition assay.

Key words Abscisic acid, PP2C, ABA receptor, ABA signaling, ABA-dependent inhibition, Phosphatase inhibition assay, pNPP

1 Introduction

Since adaptation of plants to terrestrial environments, environmental stresses such as desiccation, salinity, and cold are the major challenges to survival of plants, unlike their ancestors surrounded by aquatic environments. These environmental stresses for plants act as strong selection pressures that drive the evolution of sophisticated stress response mechanisms. A well-characterized stress response system is abscisic acid (ABA) signaling pathway responding to abiotic stresses. ABA is a phytohormone that regulates a wide range of plant processes including development, metabolism, and stress response. Abiotic stresses including drought, salinity, cold, and others induce endogenous synthesis of ABA [1]. The presence of ABA is perceived by ABA core signaling network, consisting of a ABA receptor PYRABACTIN RESISTANCE1/PYRABACTIN RESISTANCE1-LIKE/REGULATORY COMPONENTS OF ABA RECEPTOR (PYR/PYL/RCAR), a coreceptor type 2C protein

Takuya Yoshida (ed.), *Abscisic Acid: Methods and Protocols*,
Methods in Molecular Biology, vol. 2462, https://doi.org/10.1007/978-1-0716-2156-1_1,

phosphatase (PP2C), and a signal mediator sucrose nonfermenting 1-related protein kinase 2 (SnRK2) [2] (Fig. 1a, b). In the absence of ABA, PP2Cs bind to SnRK2s and subsequently inhibit their activities by blocking the catalytic site of an SnRK2, hampering autophosphorylation critical for activation of SnRK2s [3, 4] (Fig. 1a). Under stressful conditions, elevated ABA levels change the interaction network among core signaling components. PP2Cs interact with either ABA-bound ABA receptors or SnRK2s by using similar binding interfaces, but have a higher affinity toward ABA-bound ABA receptors than SnRK2s [3, 4]. Competitive interactions between PP2Cs and binding partners cause the release of SnRK2s [2, 3, 5] (Fig. 1b). As a result, SnRK2s autophosphorylate and subsequently activate numerous downstream stress response effectors by phosphorylation [6–9]. The ABA-mediated stress response allows plants to survive under harmful stress conditions, but on the other hand it has adverse effects such as growth defect. To sophisticatedly regulate the ABA response, plants have evolved highly redundant numbers of paralogs of ABA core signaling components [10–12]. Especially, both ABA receptors and PP2Cs show high redundancy and functional diversity [13, 14]. Each ABA receptor:PP2C pair exhibits different affinities toward their binding partners, thus the ABA response can be fine-tuned by a wide range of affinity strength between ABA receptor and PP2C pairs [12]. Therefore, evaluation of the functional strength of ABA receptor and PP2C complex is required to investigate the ABA-dependent regulation of ABA signaling pathway.

Protein phosphatase assay is widely used to quantify the inhibitory effect of an ABA receptor toward a PP2C. Binding of an ABA receptor to a PP2C eclipses the catalytic site of the PP2C, thereby inhibiting the enzymatic activity of the PP2C. ABA-dependent inhibition of the PP2C by the ABA receptor can be evaluated by measuring the enzymatic activities of the PP2C at different concentrations of ABA or its chemical mimics [11, 15, 16]. While there are several methods to quantify the PP2C activity, popular is a colorimetric protein phosphatase assay using pNPP (*p*-nitrophenyl phosphate). pNPP is a well-known chemical substrate to be used for a phosphatase assay. It is composed of pNP (*p*-nitrophenol) and a phosphate group. When it is hydrolyzed to pNP and inorganic phosphate by a phosphatase, the solution turns yellow due to yellow product pNP (Fig. 1c). Absorbance at 405 nm is used to measure the hydrolyzed substrate pNP. The pNPP-based colorimetric assay is convenient and fast. We will describe how to purify the recombinant proteins of the ABA receptors and PP2Cs, and how to quantify and analyze the PP2C inhibitory effects of ABA receptors in an ABA-dependent manner.

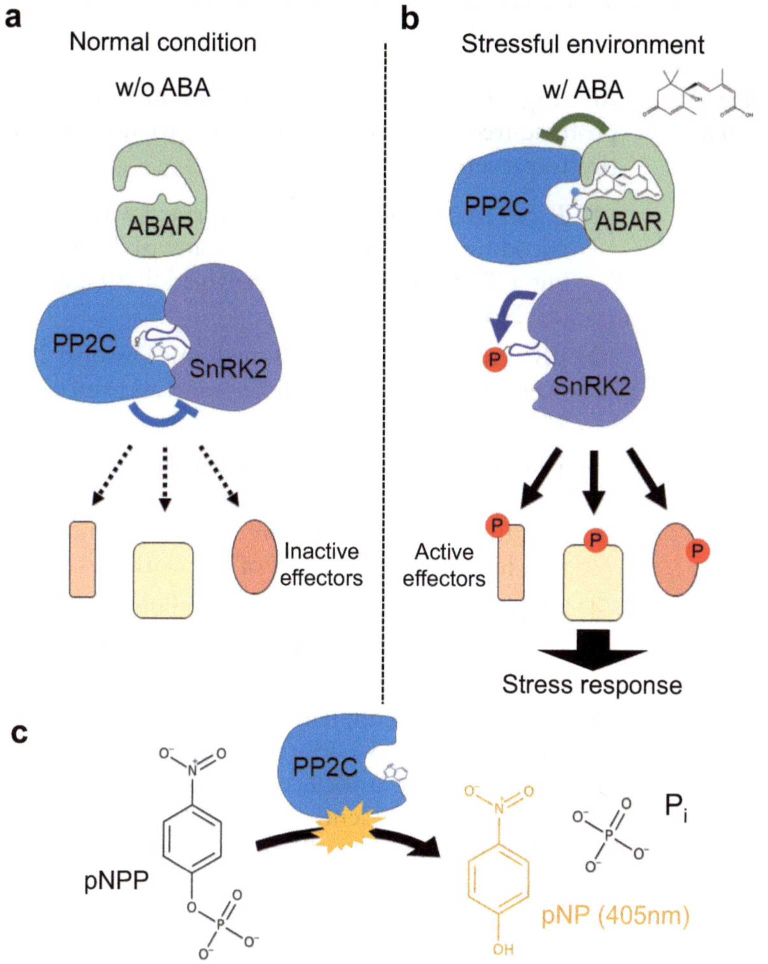

Fig. 1 Regulatory mechanism of ABA core signaling network and basic principle of protein phosphatase inhibition assay. (**a**, **b**) ABA core signaling network. The ABA core signaling network is composed of an ABA receptor (ABAR), a type 2C protein phosphatase (PP2C) and a kinase (SnRK2). (**a**) In the absence of ABA, autophosphorylation and activation of SnRK2s are inhibited by PP2Cs. (**b**) The elevation of ABA under abiotic stress is perceived by an ABAR, leading to the strong interaction between ABAR and PP2C. Since both ABAR and SnRK2 share similar interaction interfaces with PP2C, inhibition of a SnRK2 is hampered in an ABA-dependent manner. Finally, the autophosphorylated and activated SnRK2 activates various downstream effectors that mediate stress responses in plants. (**c**) The colorimetric PP2C inhibition assay based on degradation of a nonspecific substrate *p*-nitrophenyl phosphate (pNPP). pNPP can be hydrolyzed to inorganic phosphate (P_i) and *p*-nitrophenol (pNP) by PP2C. pNP assumes a yellow color which is used to quantify phosphatase activity

2 Materials

2.1 Purification of His-Tagged ABA Receptors

All reagents should be prepared in high purity, sterilized and protease-free state. We recommend to use low-binding pipette tips to prevent sample loss.

1. *E. coli* BL21 (DE3) cells transformed with a parallel-His vector encoding an ABA receptor (or any pET-based vector containing a T7 promoter and a polyhistidine-tagged ABA receptor) (*see* **Note 1**).
2. LB (Luria-Bertani) broth medium: Dissolve 20 g of LB broth (Lennox) in 1 L of deionized water and autoclave at 121 °C for 15 min.
3. 100 mg/mL ampicillin stock solution: Dissolve 1 g of ampicillin in 10 mL of deionized water. Sterilize by syringe filtration, split into aliquots and store at −20 °C until use. It is a 1000-fold higher concentration of the working concentration.
4. Ampicillin-supplemented LB medium: Before use, add ampicillin to autoclaved LB medium at a final concentration of 100 μg/mL unless otherwise noted.
5. LB agar plates supplemented with ampicillin: When you make LB medium, add 1.5% (w/v) agar powder before autoclaving. Cool down the autoclaved medium to reach below 50 °C, then add ampicillin. Gently mix for 1 min by swirling to prevent bubbles formation. Pour the medium supplemented with ampicillin to 60 mm (diameter) plates in sterile area by working near a flame.
6. 0.5 M IPTG (isopropyl β-D-1-thiogalactopyranoside) stock solution: Dissolve 1.19 g of IPTG in 10 mL of deionized water. Filter-sterilize the solution and store as small aliquots at −20 °C until use.
7. 50% (v/v) glycerol: Sterilize the pure glycerol by autoclaving at 121 °C for 15 min. Mix the same volume of glycerol and deionized water, and then vortex it until evenly mixed.
8. Flasks: Baffled flasks are recommended (*see* **Note 2**).
9. 50 and 15 mL conical tubes.
10. 1.5 mL microtubes.
11. 1 M $MgCl_2$ solution: Dissolve 0.952 g of $MgCl_2$ in 10 mL of deionized water. Sterilize the solution with syringe filter and store at room temperature (RT).
12. A spectrophotometer for measuring optical density at 600 nm wavelength (OD_{600}).
13. A tabletop centrifuge for conical tubes.

14. Buffer A: 20 mM Tris–HCl pH 8.0, 200 mM NaCl, 0.1 mM tris(2-carboxyethyl)phosphine (TCEP), 0.05% (v/v) Triton X-100.
15. Buffer B: 20 mM Tris–HCl pH 8.0, 200 mM NaCl, 0.1 mM TCEP, 20 mM imidazole.
16. Buffer C: 20 mM Tris–HCl pH 8.0, 200 mM NaCl, 0.1 mM TCEP, 200 mM imidazole.
17. Buffer D: 20 mM Tris–HCl pH 8.0, 200 mM NaCl, 0.1 mM TCEP.
18. Ni-NTA Agarose resin (e.g., Qiagen) or other immobilized metal-based affinity resins capable of binding with a polyhistidine tag.
19. A gravity flow column (e.g., Bio-Rad Glass Econo-Column 2.5 × 10 cm or equivalent).
20. A glass rod.
21. 0.5 M EDTA (ethylenediaminetetraacetic acid) pH 8.0 stock solution: Dissolve 5.84 g of EDTA in 30 mL of deionized water in a conical tube. Adjust pH to 8.0 with 5 M NaOH solution. Add deionized water up to a final volume of 40 mL. Filter-sterilize the solution and store at RT.
22. 1 L beakers.
23. A magnetic bar and a stirrer.
24. 10 kDa MWCO (molecular weight cutoff) dialysis tubing and clamps.
25. 10 kDa MWCO centrifugal filter units.
26. Ultrasonicator (e.g., Sonics Vibra-Cell).
27. A 100 mL beaker.
28. A Styrofoam® box.

2.2 Purification of GST-Tagged PP2Cs

1. *E. coli* BL21 (DE3) cells transformed with parallel GST vector encoding PP2Cs (or any expression vector containing an IPTG-inducible promoter and a GST-tagged PP2C) (*see* **Note 1**).
2. LB broth medium.
3. 100 mg/mL ampicillin stock solution.
4. Ampicillin-supplemented LB medium.
5. LB agar plates supplemented with ampicillin.
6. 0.5 M IPTG stock solution
7. 50% (v/v) glycerol
8. Flasks: Baffled flasks are recommended (*see* **Note 2**).
9. 50 and 15 mL conical tubes.

10. 1.5 mL microtubes.
11. 1 M $MgCl_2$ solution.
12. A spectrophotometer for measuring the OD_{600}.
13. A tabletop centrifuge for conical tubes.
14. Buffer A: 20 mM Tris–HCl pH 8.0, 200 mM NaCl, 0.1 mM TCEP, 0.05% (v/v) Triton X-100.
15. Buffer D: 20 mM Tris–HCl pH 8.0, 200 mM NaCl, 0.1 mM TCEP.
16. Glutathione, reduced.
17. Glutathione Sepharose 4B resin (e.g., GE Healthcare) or other glutathione affinity resins capable of binding with GST tag.
18. A gravity flow column (e.g., Bio-Rad Glass Econo-Column 2.5 × 10 cm or equivalent).
19. 1 L beakers
20. A glass rod.
21. A magnetic bar and a stirrer.
22. 10 kDa MWCO dialysis tubing and clamps.
23. 10 kDa MWCO centrifugal filter units.
24. Ultrasonicator (e.g., Sonics Vibra-Cell).
25. A 100 mL beaker.
26. A Styrofoam® box.

2.3 Protein Phosphatase Inhibition Assay

1. A 96-well microplate reader capable of measuring absorbance at 405 nm wavelength (A_{405}).
2. 96-well microplates (clear, flat-bottom, nontreated).
3. 10× phosphatase reaction buffer: 20 mM Tris–HCl pH 8.0, 10 mM $MnCl_2$, 1% (v/v) β-mercaptoethanol, 0.1% (w/v) BSA. To prevent oxidation of β-mercaptoethanol, make the buffer just before use.
4. 100 mM ABA stock solution: Dissolve 26.4 mg of ABA in 1 mL of DMSO (Dimethyl sulfoxide) and store at −20 °C.
5. 10× (150 mM) *p*-nitrophenyl phosphate (pNPP) solution: Dissolve 0.278 g of *p*-nitrophenyl phosphate disodium salt hexahydrate in 5 mL of deionized water. To avoid degradation of pNPP, make the solution immediately before use.
6. Purified ABA receptors and PP2Cs (Subheadings 3.1 and 3.2): Store at −80 °C until use.
7. Multichannel pipettes covering the volume range from 10 to 80 μL.
8. 25 mL disposable reagent reservoirs.
9. A tabletop refrigerated centrifuge for 1.5 mL microtubes.

10. A tabletop centrifuge for 96-well plates.
11. An orbital shaker.
12. Bradford assay reagents and a spectrophotometer capable of measuring absorbance at 595 nm wavelength.
13. ImageJ or other image analysis software for analysis of protein bands on SDS-PAGE.
14. Statistical analysis software (e.g., GraphPad Prism or equivalent).
15. Microplate sealing films (adhesive, clear).
16. A stationary incubator for maintaining a temperature of 37 °C.
17. 15 mL conical tubes.
18. 1.5 mL microtubes.

3 Methods

3.1 Purification of His-Tagged ABA Receptors

1. Pick a single colony from an ampicillin-supplemented LB agar plate of *E. coli* BL21 (DE3) cells transformed with a parallel-His vector encoding an ABA receptor and preculture the cells in 10 mL of LB medium supplemented with ampicillin at 37 °C overnight, 200 rpm.
2. Transfer precultured cells to 0.5 L of LB medium supplemented with ampicillin, 2 mM $MgCl_2$, and 0.5% (v/v) glycerol in a 2 L baffled flask. Keep your bench area sterile by working near a flame.
3. Grow cells by shaking at 37 °C and 120 rpm until OD_{600} reaches 0.6–1.0 (*see* **Note 3**).
4. Induce protein expression by addition of IPTG to a final concentration of 0.1 mM, and then incubate for 5 h or desired time period at 16 °C and 120 rpm (*see* **Note 4**).
5. Harvest cells expressing a His-tagged ABA receptor by centrifugation at 3300 × *g* for 15 min, and then discard the supernatant. You can freeze and store the harvested cells at −80 °C until use.
6. Resuspend cells in 40 mL of the buffer A.
7. Lyse resuspended cells by ultrasonication on a ultrasonicator with 2 s of pulses and 2 s of intervals, until it becomes clear. During sonication, the lysate should be kept at cold temperature to prevent loss (*see* **Note 5**).
8. Pour the lysate to a conical tube and centrifuge it to remove cell debris at 16,000 × *g* for 30 min at 4 °C.
9. During centrifugation, equilibrate 5 mL of Ni-NTA Agarose resin in a gravity-flow column with the buffer B (*see* **Note 6**).

Wash the column by pouring 10 bed volumes of deionized water, and then wash with 10 bed volumes of the buffer B. Be aware not to make the resin dry during purification process (*see* **Note 7**).

10. Collect the supernatant of the centrifuged lysate and the supernatant should be kept on ice. Close a stopcock valve of the column and load the supernatant to the Ni-NTA Agarose-filled column (*see* **step 9**). Make sure the pellet does not flow into the column.
11. Open a valve on the column and allow the His-tagged ABA receptor to bind with the Ni-NTA agarose resin by passing the solution through the column.
12. Wash the resin with 30 bed volumes of the buffer B. You should be careful not to float the resin while you pour the buffer.
13. Elute the protein by passing 3–5 bed volumes of the buffer C. After elution, immediately add 0.5 M EDTA stock solution to a final concentration of 0.5 mM (*see* **Note 8**).
14. Dialyze the eluate against 1 L of the buffer D in 1 L beaker by stirring with a 10 kDa MWCO dialysis tubing. Leave it at 4 °C for overnight.
15. Concentrate the purified His-tagged ABA receptor using a 10 kDa MWCO centrifugal filter unit, until a desired concentration is reached (*see* **Note 9**).
16. Add glycerol at a final concentration of 10% (v/v) to the dialyzed protein. Gently mix the protein and the glycerol.
17. Freeze the protein and the glycerol mixture in liquid nitrogen, and store at −80 °C.

3.2 Purification of GST-Tagged PP2Cs

1. Pick a single colony from an ampicillin-supplemented LB agar plate of *E. coli* BL21 (DE3) cells transformed with parallel-GST vector encoding a PP2C and preculture the cells in 10 mL of LB medium supplemented with ampicillin at 37 °C overnight, 200 rpm.
2. Transfer precultured cells to 0.5 L of LB medium supplemented with ampicillin, 2 mM $MgCl_2$, and 0.5% (v/v) glycerol in a 2 L baffled flask. Keep your bench area sterile by working near a flame.
3. Grow cells by shaking at 37 °C and 120 rpm until OD_{600} reaches 0.6–1.0 (*see* **Note 3**).
4. Induce protein expression by addition of IPTG to a final concentration of 0.1 mM, and then incubate for overnight or desired time at 16 °C and 120 rpm.

5. Harvest cells expressing the GST-tagged PP2C by centrifugation at 3300 × *g* for 15 min, and then discard supernatant. You can freeze and store the harvested cells at −80 °C until use.
6. Resuspend cells in 40 mL of the buffer A.
7. Lyse resuspended cells by ultrasonication on a ultrasonicator with 2 s of pulses and 2 s of intervals, until it becomes clear. During sonication, the lysate should be kept at cold temperature to prevent loss (*see* **Note 5**).
8. Pour the lysate to conical tube and centrifuge it to remove cell debris at 16,000 × *g* for 30 min at 4 °C.
9. During centrifugation, equilibrate 5 mL of glutathione Sepharose 4B resin in a gravity-flow column with the buffer D (*see* **Note 6**). Wash the column with 10 bed volumes of deionized water, and then wash with 10 bed volumes of the buffer D. Be aware not to make the resin dry during purification process (*see* **Note 7**).
10. Collect the supernatant of the centrifuged lysate. The supernatant should be kept on ice. Close the valve and load the supernatant to the glutathione Sepharose 4B-filled column (*see* **step 9**). Make sure the pellet does not flow into the column.
11. Using a glass rod, stir the resin and the lysate supernatant every few minutes. After 30 min, open the valve and pass the lysate through resin.
12. Wash the resin with 30 bed volumes of the buffer D. You should be careful not to float the resin while you pour the buffer.
13. Close the column valve, and then stir with 1 bed volume of the buffer D with 10 mM reduced glutathione (*see* **Note 10**). Wait for 5 min to elute out and collect it. Repeat this step three to five times. Repeated application of elution buffer can increase the total yield.
14. Dialyze the eluate against 1 L of the buffer D in 1 L beaker by stirring with a 10 kDa MWCO dialysis tubing. Leave it at 4 °C for overnight.
15. Concentrate the purified GST-tagged PP2Cs using a 10 kDa MWCO centrifugal filter unit, until a desired concentration is reached (*see* **Note 9**).
16. Add glycerol at a final concentration of 10% (v/v) to the dialyzed protein. Gently mix the protein and glycerol.
17. Freeze the protein and glycerol mixture in liquid nitrogen, and store at −80 °C.

17. While waiting for the incubation of protein mixtures, prepare the 10× pNPP solution.
18. Pour the 10× pNPP solution into the reagent reservoir.
19. Gently detach the sealing film from the microplate after incubation, and add 10 μL of the 10× pNPP solution to wells on the columns 2–4 (A2 through H4) using a multichannel pipette (Fig. 2h) (*see* **Note 14**).
20. Seal the microplate again using a new sealing film.
21. Spin down the microplate at 200 × *g* for 2 min.
22. Shake the microplate at 200 rpm for 1 min.
23. Initiate protein phosphatase reaction by incubating the microplate at 37 °C for 1 h (*see* **Note 15**).
24. Spin down the microplate at 200 × *g* for 2 min and detach the sealing film gently (*see* **Note 16**).
25. Measure absorbance at 405 nm wavelength of the wells on the columns 2–4 (A2 through H4) with a microplate reader (Fig. 2i).
26. Use an appropriate statistical analysis software to quantify the half maximal inhibitory concentration (IC_{50}) of ABA using the equation shown in Fig. 3a (*see* **Note 17**).

a

$$\log_{10}[ABA] = \frac{100}{1 + 10^{RA_{PP2C} - \log IC_{50}}}$$

b

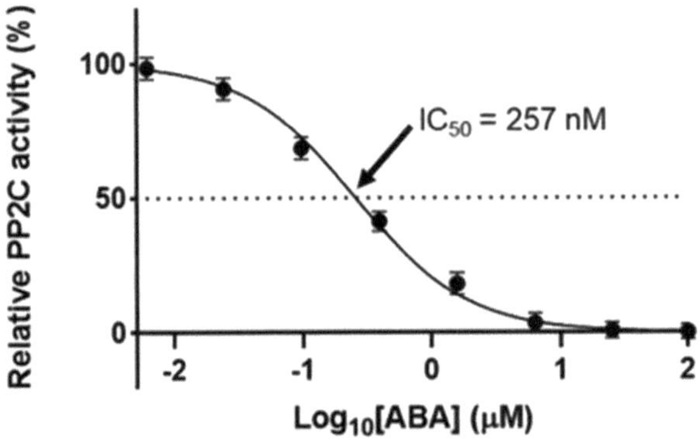

Fig. 3 Calculation of the half maximal inhibitory concentration (IC_{50}) of ABA. (**a**) Equation to calculate IC_{50} values. RA_{PP2C} refers to relative activity of PP2C (%) and log[ABA] is shown in the unit of μM. (**b**) An exemplary graph in which the curve was fitted to data points of PP2C inhibition assay using the equation in (A). Graph and IC_{50} value were calculated by Prism (GraphPad) software

4 Notes

1. Codon optimization of a gene encoding a protein is recommended due to differences in codon usage between *E. coli* and plants. In addition, rare codon tRNAs-supplemented *E. coli* strains such as Rosetta 2 (DE3) are also recommended to use for increasing protein yield. Refer to [17].
2. Baffled flasks are important for oxygen supply into medium. If you use nonbaffled flat flasks, protein expression level will be lower than baffled flasks. Usage of nonbaffled flat flasks is also acceptable, but not recommended.
3. Since doubling time of *E. coli* is affected by various factors, you should check the optical density every half-hour after 1 h from inoculation.
4. For some ABA receptors or their mutants, overnight expression causes protein degradation, lowering the protein yield. We recommend to check at least two conditions: 5 h and 12 h expression. If the desired ABA receptor seems to be unstable upon expression, supplementation of the LB medium with a final concentration of 10 μM ABA will help the ABA receptor expression.
5. The lysate should be kept cold during sonication because heat generated from sonication process can cause irreversible damage to proteins. We recommend to place an 100 mL glass beaker on a Styrofoam® box, and then fill the space between the beaker and the box with ice and cold water to maximize thermal conductivity. Fill the beaker with resuspended cells and sonicate the cells.
6. If you pour a buffer directly to affinity chromatography resin in a column, the resin particles will float. Therefore, put a glass rod on the wall inside the column, and pour the buffer on the tilted glass rod. Then, the resin would settle on the bottom.
7. Drying affinity chromatography resin may cause oxidation of a protein depending on a protein of interest. PP2Cs are known to be oxidation-sensitive and their phosphatase activities are inhibited by oxidation [18–20].
8. The elution buffer causes leakage of nickel ions from Ni-NTA resin, and free nickel ions in the solution lead to aggregation of His-tagged proteins. To prevent aggregation, we usually remove the free nickel ions in the eluate by adding 0.5 mM EDTA. EDTA will capture the free nickel ions in the solution, and then it will be removed by dialysis in a later step.
9. In consideration of each component volume of phosphatase assay, protein concentrations higher than 10 μM are recommended.

10. Only reduced form of glutathione can be recognized by GST and elute the GST tagged protein. Therefore, immediately add reduced glutathione to the elution buffer before use.

11. Vigorously vortex for at least 30 s to evenly dilute the ABA solution.

12. Generally, purified proteins from affinity chromatography contain small amounts of contaminants that have a negligible effect on assay. However, some ABA receptors or PP2Cs show low expression level, and purified proteins contain large amounts of contaminants after enrichment. To determine accurate concentrations of target protein, you should check the purity of protein using SDS-PAGE after protein purification, and calculate the relative proportion of a target protein by ImageJ or other image analysis software. After Bradford assay, a target protein concentration can be determined by multiplying the relative proportion of the target protein by the total protein concentration from Bradford assay.

13. The protein reaction mixture is prepared as follows. First of all, determine the volume of the purified PP2C and the purified ABA receptor to make the final concentration in the mixture be 100 nM and 500 nM, respectively. Note that the concentration of the purified PP2C or ABA receptor varies from batch to batch. If we set the volume determined for the deionized water as w (in μL), that for the purified PP2C as p (in μL) and that for the purified ABA receptor as a (in μL), the relationship among volumes of the components in the protein reaction mixture is

$$3000 = 375 + w + p + a$$

where the number 3000 refers to the total volume of the protein reaction mixture in μL. Add 375 μL of the phosphatase reaction buffer and the w μL of the deionized water in a 15 mL conical tube, and mix the solution by gently inverting the tube 10–20 times. Then, add p μL of the purified PP2C and a μL of the purified ABA receptor into the tube containing the phosphatase buffer and the deionized water, and gently invert 20–30 times to mix evenly.

14. In the final step of assay, each well on the microplate contains 80 μL of the 1.25× protein reaction mixture, 10 μL of the 10× ABA working solution, and 10 μL of 10× pNPP solution, bringing the total volume of 100 μL. The final concentrations of the purified PP2C, the purified ABA receptor and the phosphatase reaction buffer are 80 nM, 400 nM, and 1×, respectively.

15. Incubation time can be adjusted depending on activity of PP2C, but keep in mind that pNPP is a very unstable chemical.

16. If you want to stop the phosphatase reaction, add 50 μL of 2 M NaOH solution to each well, then spin down and shake as described above.
17. After assay, draw the dose-response curve to calculate IC_{50} (Fig. 3b). We should convert the results to relative activity value. Statistical analysis software (e.g., GraphPad Prims) can estimate bottom and top plateaus from nonlinear regression analysis. Relative protein phosphatase activity will be 0% at the bottom plateaus and 100% at the top plateaus. However, if IC_{50} value of your protein is out of range, then estimation of plateaus will be inaccurate. In that case, you should change the concentrations of ABA serial dilution solution and repeat the assay.

References

1. Xiong L, Zhu JK (2003) Regulation of abscisic acid biosynthesis. Plant Physiol 133(1):29–36. https://doi.org/10.1104/pp.103.025395
2. Fujii H, Chinnusamy V, Rodrigues A, Rubio S, Antoni R, Park SY, Cutler SR, Sheen J, Rodriguez PL, Zhu JK (2009) In vitro reconstitution of an abscisic acid signalling pathway. Nature 462(7273):660–664. https://doi.org/10.1038/nature08599
3. Soon F-F, Ng L-M, Zhou XE, West GM, Kovach A, Tan MHE, Suino-Powell KM, He Y, Xu Y, Chalmers MJ, Brunzelle JS, Zhang H, Yang H, Jiang H, Li J, Yong E-L, Cutler S, Zhu J-K, Griffin PR, Melcher K, Xu HE (2012) Molecular mimicry regulates ABA signaling by SnRK2 kinases and PP2C phosphatases. Science 335(6064):85–88. https://doi.org/10.1126/science.1215106
4. West GM, Pascal BD, Ng LM, Soon FF, Melcher K, Xu HE, Chalmers MJ, Griffin PR (2013) Protein conformation ensembles monitored by HDX reveal a structural rationale for abscisic acid signaling protein affinities and activities. Structure 21(2):229–235. https://doi.org/10.1016/j.str.2012.12.001
5. Brandt B, Brodsky DE, Xue S, Negi J, Iba K, Kangasjarvi J, Ghassemian M, Stephan AB, Hu H, Schroeder JI (2012) Reconstitution of abscisic acid activation of SLAC1 anion channel by CPK6 and OST1 kinases and branched ABI1 PP2C phosphatase action. Proc Natl Acad Sci U S A 109(26):10593–10598. https://doi.org/10.1073/pnas.1116590109
6. Belin C, de Franco PO, Bourbousse C, Chaignepain S, Schmitter JM, Vavasseur A, Giraudat J, Barbier-Brygoo H, Thomine S (2006) Identification of features regulating OST1 kinase activity and OST1 function in guard cells. Plant Physiol 141(4):1316–1327. https://doi.org/10.1104/pp.106.079327
7. Takahashi Y, Ebisu Y, Shimazaki KI (2017) Reconstitution of abscisic acid signaling from the receptor to DNA via bHLH transcription factors. Plant Physiol 174(2):815–822. https://doi.org/10.1104/pp.16.01825
8. Fujita Y, Nakashima K, Yoshida T, Katagiri T, Kidokoro S, Kanamori N, Umezawa T, Fujita M, Maruyama K, Ishiyama K, Kobayashi M, Nakasone S, Yamada K, Ito T, Shinozaki K, Yamaguchi-Shinozaki K (2009) Three SnRK2 protein kinases are the main positive regulators of abscisic acid signaling in response to water stress in Arabidopsis. Plant Cell Physiol 50(12):2123–2132. https://doi.org/10.1093/pcp/pcp147
9. Fujii H, Zhu J-K (2009) Arabidopsis mutant deficient in 3 abscisic acid-activated protein kinases reveals critical roles in growth, reproduction, and stress. Proc Natl Acad Sci U S A 106(20):8380. https://doi.org/10.1073/pnas.0903144106
10. Ma Y, Szostkiewicz I, Korte A, Moes D, Yang Y, Christmann A, Grill E (2009) Regulators of PP2C phosphatase activity function as abscisic acid sensors. Science 324(5930):1064. https://doi.org/10.1126/science.1172408
11. Park SY, Fung P, Nishimura N, Jensen DR, Fujii H, Zhao Y, Lumba S, Santiago J, Rodrigues A, Chow TF, Alfred SE, Bonetta D, Finkelstein R, Provart NJ, Desveaux D, Rodriguez PL, McCourt P, Zhu JK, Schroeder JI, Volkman BF, Cutler SR (2009) Abscisic acid inhibits type 2C protein phosphatases via the PYR/PYL family of START proteins. Science 324(5930):

1068–1071. https://doi.org/10.1126/science.1173041

12. Tischer SV, Wunschel C, Papacek M, Kleigrewe K, Hofmann T, Christmann A, Grill E (2017) Combinatorial interaction network of abscisic acid receptors and coreceptors from Arabidopsis thaliana. Proc Natl Acad Sci U S A 114(38):10280–10285. https://doi.org/10.1073/pnas.1706593114
13. Singh A, Pandey A, Srivastava AK, Tran LP, Pandey GK (2016) Plant protein phosphatases 2C: from genomic diversity to functional multiplicity and importance in stress management. Crit Rev Biotechnol 36(6):1023–1035. https://doi.org/10.3109/07388551.2015.1083941
14. Miyakawa T, Fujita Y, Yamaguchi-Shinozaki K, Tanokura M (2013) Structure and function of abscisic acid receptors. Trends Plant Sci 18(5): 259–266. https://doi.org/10.1016/j.tplants.2012.11.002
15. Hao Q, Yin P, Yan C, Yuan X, Li W, Zhang Z, Liu L, Wang J, Yan N (2010) Functional mechanism of the abscisic acid agonist pyrabactin. J Biol Chem 285(37):28946–28952. https://doi.org/10.1074/jbc.M110.149005
16. Melcher K, Xu Y, Ng LM, Zhou XE, Soon FF, Chinnusamy V, Suino-Powell KM, Kovach A, Tham FS, Cutler SR, Li J, Yong EL, Zhu JK, Xu HE (2010) Identification and mechanism of ABA receptor antagonism. Nat Struct Mol Biol 17(9):1102–1108. https://doi.org/10.1038/nsmb.1887
17. Han S, Min MK, Lee S-Y, Lim CW, Bhatnagar N, Lee Y, Shin D, Chung KY, Lee SC, Kim B-G, Lee S (2017) Modulation of ABA signaling by altering VxGΦL motif of PP2Cs in Oryza sativa. Mol Plant 10(9): 1190–1205. https://doi.org/10.1016/j.molp.2017.08.003
18. Meinhard M, Grill E (2001) Hydrogen peroxide is a regulator of ABI1, a protein phosphatase 2C from Arabidopsis. FEBS Lett 508(3): 443–446. https://doi.org/10.1016/S0014-5793(01)03106-4
19. Meinhard M, Rodriguez PL, Grill E (2002) The sensitivity of ABI2 to hydrogen peroxide links the abscisic acid-response regulator to redox signalling. Planta 214(5):775–782. https://doi.org/10.1007/s00425-001-0675-3
20. Ni L, Fu X, Zhang H, Li X, Cai X, Zhang P, Liu L, Wang Q, Sun M, Wang Q-W, Zhang A, Zhang Z, Jiang M (2019) Abscisic acid inhibits rice protein phosphatase PP45 via H_2O_2 and relieves repression of the Ca^{2+}/CaM-dependent protein kinase DMI3. Plant Cell 31(1): 128. https://doi.org/10.1105/tpc.18.00506

Chapter 2

Inhibition of SnRK2 Kinases by Type 2C Protein Phosphatases

Ewa Krzywinska, Katarzyna Patrycja Szymanska, and Grazyna Dobrowolska

Abstract

SNF1-related protein kinase 2 s (SnRK2s) are major regulators of plant growth, development and responses to environmental stresses. Together with clade A protein phosphatases of type 2C (PP2C) and REGU LATORY COMPONENTS OF ABA RECEPTOR (RCAR also known as PYRABACTIN RESISTANCE1 (PYR1) or PYR1-LIKE (PYL)) soluble abscisic acid (ABA) receptors they form the core of ABA-signaling. Clade A PP2Cs play a negative role in ABA signaling, primarily by inhibiting SnRK2 activity, through direct interaction and dephosphorylation of SnRK2s. Here, we describe two methods, which can be used for monitoring inhibition of the SnRK2 activity by PP2C phosphatases. One of them is an in vitro dephosphorylation assay using SnRK2 as the substrate followed by a classical in-gel kinase-activity assay and the other is immunocomplex kinase-activity assay, which can be applied for analysis of the SnRK2 activity in plant material.

Key words SnRK2, PP2C, Protein phosphatase, Dephosphorylation assay, Abscisic acid, Kinase activity

1 Introduction

The largest protein phosphatase family in plants is protein phosphatases of type 2C (PP2C), with 80 members in Arabidopsis and 90 in rice, classified into 12 clades (A-K) and 11 clades (A-J), respectively, based on phylogenetic analyses [1]. These phosphatases are Mg^{2+} or Mn^{2+}-dependent monomeric enzymes, which consist of a catalytic domain, most often situated at the C-terminus, and N-terminal or C-terminal extensions bearing distinct additional domains depending on the phosphatase [2]. It is well established that nine clade A and three clade E PP2C phosphatases are negative regulators of stress responses and abscisic acid (ABA) signaling [for recent reviews see: [3–5] and references herein]. They dephosphorylate numerous proteins involved in the

Takuya Yoshida (ed.), *Abscisic Acid: Methods and Protocols*,
Methods in Molecular Biology, vol. 2462, https://doi.org/10.1007/978-1-0716-2156-1_2,

ABA-dependent as well as the ABA-independent responses, being a hub in the crosstalk between various signaling pathways. The best known targets of clade A PP2C phosphatases are SNF1-related protein kinase 2 s (SnRK2s), which are considered as key regulators of plant responses to environmental stresses, development, and even growth in nonstress conditions. Among the identified SnRK2 substrates are ABA-responsive transcription factors, ion channels controlling ABA-dependent stomatal movements, proteins involved in regulation of microRNA biogenesis, flowering and several others [for recent reviews see: [4–8]]. In the past two decades, substantial progress has been made in understanding how ABA is sensed inside the cell and which proteins are involved in ABA signal transduction. A breakthrough in understanding the mechanism of ABA signal transduction was in the year 2009, when the REGULATORY COMPONENTS OF ABA RECEPTOR (RCAR also known as PYRABACTIN RESISTANCE1 (PYR1) or PYR1-LIKE (PYL)) proteins were identified as soluble ABA receptors [9, 10]. Furthermore, it has been shown that, together with the clade A PP2Cs (named coreceptors) [9, 10] and ABA-responsive SnRK2s, they form a core of ABA-signaling [11, 12]. The knowledge on these elements and their relationship gave us a sense of the first stages in ABA signaling. In short, upon ABA binding, the RCAR receptors bind PP2Cs and inhibit their activity allowing SnRK2s activation and phosphorylation of their targets [8, 11, 12]. On the other hand, in optimal growth conditions, PP2Cs inhibit SnRK2s by direct interaction and dephosphorylation [13–15]. This is an obvious simplification of the mechanism of ABA signaling, which in nature is much more complicated. There are several other proteins involved in ABA signaling triggered by developmental cues or stress conditions. All elements of the core of ABA signalosome (ABA receptors, PP2Cs and SnRK2s) are highly regulated by interaction with other proteins or second messengers, posttranslational modifications, and degradation. Each year several new data concerning their regulators are published. Very recently this subject has been covered by excellent reviews [4–6]. It should also be noted that some of clade A PP2Cs (namely, ABA INSENSITIVE 1 (ABI1) and ABA-HYPERSENSITIVE GE RMINATION 3 (AHG3), also known as PP2CA) besides the ABA-responsive SnRK2s inhibit SnRK2.4 (also known as SRK2A), which is not activated by treatment of plant material with ABA, but by, for example, osmotic stress [16, 17]. Here, we describe both an in vitro as well as an in vivo method to analyze the impact of PP2C phosphatases on SnRK2 kinases, with ABI1 as an example. The in vitro assay is based on using recombinant SnRK2 kinases as substrates for the PP2C phosphatase in dephosphorylation reactions. This is followed by analysis of the kinase activity using the in-gel kinase assay method, rather than a simple in vitro kinase assay. Electrophoretic separation of the phosphatase and

SnRK2 kinases ensures that the kinase substrate myelin basic protein (MBP), crosslinked in the gel, is not in parallel dephosphorylated by the phosphatase. The other method, immunocomplex kinase activity assay, allows for analysis of the activity of SnRK2 kinases from plant material. Here, as an example SnRK2 activity was monitored in Arabidopsis T87 cells and T87 cells ectopically expressing epitope-tagged ABI1, after exposure to salt stress. This method could be utilized to monitor SnRK2 activity, for example, in Arabidopsis mutants, coexpression and overexpression studies or after plants/cell exposure to specific low-molecular weight regulators of the plant signaling pathways.

2 Materials

2.1 Purification of Recombinant Proteins

1. *SnRK2* cloned into a pGEX vector with a thrombin cleavage site and transformed into *E. coli* BL21 (DE3) bacteria.
2. *PP2C* cloned into a pGEX vector and transformed into *E. coli Rosetta* (DE3) (*see* **Note 1**).
3. LB medium.
4. 100 mg/mL ampicillin.
5. 0.1 mg/mL IPTG.
6. Resuspension buffer: 20 mM Tris–HCl (pH 8.8), 500 mM NaCl, 1 mM DTT (*see* **Note 2**).
7. Protein purification buffer: 20 mM Tris–HCl (pH 7.5), 150 mM NaCl, 1 mM DTT (*see* **Note 2**).
8. 0.1 mg/mL lysozyme, freshly prepared.
9. Viscolase (A&A Biotechnology, 250 U/μL; *see* **Note 3**).
10. Glutathione-Sepharose 4B resin.
11. Thrombin (1 U/100 μL of resin).
12. STE buffer: 10 mM Tris–HCl (pH 8.0), 1 mM EDTA, 150 mM NaCl.
13. 10% sarkosyl (sodium lauroyl sarcosinate).
14. 1 M DTT.
15. 10% Triton X-100.
16. 20 mM Tris–HCl, pH 8.0.
17. 250 mM reduced glutathione (GSH).
18. Bradford reagent.
19. Spectrophotometer for measurement of bacterial culture optical density and determining the absorbance of samples in Bradford protein assay.

20. Commercial kit for assaying serine/threonine phosphatase activity (e.g., Promega Serine/Threonine Phosphatase Assay System or ProFluor® Ser/Thr PPase Assay).
21. 5× phosphatase buffer: 100 mM Tris–HCl (pH 7.5), 125 mM $MgCl_2$.
22. 50 mL conical tube.
23. 2 mL Eppendorf tube.
24. Rotating tube mixer.
25. Mini spin column.
26. Ultrasonic homogenizer (e.g., Branson Ultrasonics Corp.).

2.2 In Vitro Phosphatase Assay

1. 500 mM $MgCl_2$.
2. 100 mM Tris–HCl (pH 7.5).
3. 5× SDS sample buffer: 250 mM Tris–HCl (pH 6.8), 10% SDS, 50% glycerol, 0.5 M DTT, 0.25% bromophenol blue.
4. 1.5 mL tube.

2.3 In-Gel Kinase Assay

1. SDS Removal Solution (SRS): 25 mM Tris–HCl (pH 7.5), 5 mM NaF, 0.1 mM Na_3VO_4, 0.05% BSA, 0.1% Triton X-100, 0.1 mM DTT (*see* **Note 2**).
2. Renaturation Solution: 25 mM Tris–HCl (pH 7.5), 5 mM NaF, 0.1 mM Na_3VO_4, 0.1 mM DTT (*see* **Note 2**).
3. Kinase Reaction Solution (KRS): 40 mM HEPES (pH 7.5), 13.5 mM $MgCl_2$, 2 mM EGTA (pH 7.0,) 0.1 mM Na_3VO_4, 0.1 mM DTT (*see* **Note 2**).
4. Gel Wash Solution (GWS): 5% TCA, 70 mM NaPPi.
5. Preservation Solution: 50% methanol, 5% glycerol.
6. Resolving gel mix (5 mL/gel): 350 mM Tris–HCl (pH 8.8), 12% acrylamide/bisacrylamide, 0.5 mg/mL MBP, 0.1% SDS, 0.1% Ammonium persulfate (APS), 0.01% TEMED.
7. Stacking gel mix (3 mL/gel): 200 mM Tris–HCl (pH 6.8), 4% acrylamide/bisacrylamide, 0.1% SDS, 0.1% APS, 0.01% TEMED.
8. Isopropanol.
9. 10× Tris-Glycine SDS-PAGE running buffer: Dissolve 30.3 g Tris, 144.0 g glycine, and 10.0 g SDS in distilled water to a final volume of 1 L. Use as 1× running buffer by diluting 100 mL of the 10× running buffer to 1 L with distilled water.
10. Prestained protein ladder.
11. [γ-^{32}P] ATP (10 mCi/mL).
12. 1 mM ATP.
13. Coomassie BB Staining Solution.

14. Plastic container with lid for incubation of polyacrylamide gels.
15. Gel drying film.
16. Orbital platform with gentle motion.
17. Basic equipment needed for protein electrophoresis.
18. Equipment for visualization of radiolabeled proteins – phosphor imaging system or equipment for autoradiography (X-ray film and appropriate metal film cassette, developer, fixer, trays for film development, clips for hanging of film).

2.4 Immunocomplex Kinase Activity Assay

1. Plant material (any type of plant material in which one would like to assay SnRK2 kinase activity).
2. Immunoprecipitation buffer: 20 mM Tris–HCl (pH 7.5), 2 mM EDTA, 2 mM EGTA, 1% Triton X-100, 150 mM NaCl, 50 mM β-glycerophosphate, 10 mM Na_3VO_4, 10 mM DTT, 1 mM PMSF, 1× Protease Inhibitor Cocktail (*see* **Note 4**).
3. Anti-SnRK2 antibody (*see* **Note 5**); in this protocol custom-made rabbit polyclonal anti-SnRK2.4/SnRK2.10 (also known as SRK2B) antibodies raised against the N-terminal peptide (MDKYELVKDIGC) of these kinases.
4. Protein-A agarose (*see* **Note 6**).
5. Sonicator for multiple samples (e.g., Diagenode Bioruptor Plus or equivalent).
6. Benchtop mini centrifuge.
7. Rotating tube mixer.

3 Methods

3.1 In Vitro Assay for Determining the Effect of PP2C Phosphatase on SnRK2 Kinases

3.1.1 Purification of Recombinant SnRK2 Kinase

1. Prepare overnight culture of *E. coli* BL21(DE3) by inoculation of a single colony of bacteria carrying an appropriate plasmid into 5 mL of LB medium supplemented with 5 μL of ampicillin (final conc. 100 μg/mL) and incubate at 37 °C with shaking 180 rpm. Seed the bacteria in a 1:100 ratio into 500 mL of LB medium supplemented with 500 μL of ampicillin (100 μg/mL) and incubate at 37 °C with shaking 180 rpm until OD_{600} reaches 0.6–0.7.
2. Induce the recombinant protein synthesis by addition of IPTG into LB medium to a final concentration of 0.8 mM and incubation at 18 °C with shaking 180 rpm overnight. Centrifuge the overnight cultures at 4000 × *g* 4 °C for 10 min, discard the supernatant and collect the bacterial pellet into a 50 mL conical tube for further manipulations (*see* **Note 7**).

3. Resuspend the bacterial pellet in 20 mL of the resuspension buffer (previously supplemented with DTT; *see* **Note 2**) and incubate it with 200 μL of 0.1 mg/mL lysozyme and Viscolase (1.5 μL, *see* **Note 3**) on a rotating tube mixer for 30 min at 4 °C.
4. Sonicate the lysates on ice 3 times for 1 min with 1 min interval (*see* **Note 8**) and centrifuge at 10,000 × *g*, 4 °C for 20 min. In the meantime, pipet 0.5 mL of glutathione-Sepharose 4B resin into a 50 mL conical tube, add 20 mL of protein purification buffer (supplemented with DTT) and centrifuge at 500 × *g* for 5 min. Discard the supernatant.
5. After the centrifugation of the bacterial lysate is complete, keep a small portion of the supernatant for SDS-PAGE analysis (30 μL, *see* **Note 9**) and add the rest into the tube with equilibrated resin. Incubate on a rotating tube mixer for 2 h at 4 °C.
6. Centrifuge the resin at 500 × *g* for 5 min, carefully discard supernatant, saving a small portion for further analysis (*see* **Note 9**) and resuspend resin in 10 mL of the resuspension buffer. Repeat this step 7 times and additionally 2 times using the protein purification buffer (*see* **Note 10**). After the final wash resuspend the resin in 700 μL of the protein purification buffer and transfer it into a 2 mL Eppendorf tube (*see* **Note 11**).
7. To perform protein cleavage, add thrombin (1 U/100 μL of resin initial volume) into the resin suspension and incubate it on a rotating tube mixer overnight at 4 °C. Then transfer the suspension into a mini spin column and centrifuge for 4 min at 1000 × *g* at 4 °C. Collect the eluate and determine the protein concentration, for example, using Bradford protein assay. Aliquot and store at −80 °C. Additionally, run SDS-PAGE gel to determine the purified protein quality, together with a sample of the resuspended resin after cleavage with thrombin and the samples from previous steps (*see* **Note 9**).

3.1.2 Purification of Recombinant GST-Fused PP2C Phosphatase

The following protocol is a modified version of a method described by Frangioni and Neel [18].

1. Prepare overnight bacterial cultures by inoculating 5 mL of LB medium supplemented with 100 μg/μL ampicillin with a single colony of bacteria carrying the pGEX-PP2C or empty pGEX plasmid, grow overnight at 37 °C with shaking. Seed the overnight culture into 500 mL of LB medium with ampicillin (100 μg/mL), and grow at 37 °C until OD_{600} reaches 0.5.
2. Induce recombinant protein synthesis by adding IPTG to a final concentration of 0.8 mM and incubate for 2 h at 37 °C. Centrifuge the culture at 4000 × *g*, 4 °C for 10 min.

3. Decant supernatant and resuspend pellet on ice in 10 mL of cold STE buffer. Add 100 μL of freshly prepared lysozyme solution and incubate for 15 min on ice. Add 100 μL of 1 M DTT and 1.4 mL of 10% sarkosyl, mix by inversion.
4. Sonicate two times for 30 s with a 30 s interval. Centrifuge at 10,000 × *g*, 4 °C for 20 min.
5. Transfer the supernatant to a 50 mL conical tube add 4 mL of 10% Triton X-100 and top up with STE buffer to 20 mL. Incubate for 30 min at room temperature (RT).
6. Pipet 1 mL of glutathione-Sepharose resin into a new conical tube, add 20 mL of STE buffer, mix by inversion and centrifuge for 500 × *g* for 5 min. Carefully remove supernatant. Pour the bacterial lysate (*see* **Note 9**) into the prepared resin and incubate at 4 °C for 2 h on a rotating tube mixer.
7. Centrifuge for 500 × *g* for 5 min, carefully remove supernatant (saving a small fraction, *see* **Note 9**) and add 20 mL of STE buffer, incubate at 4 °C for 10 min on a rotating tube mixer. Centrifuge as before and repeat the wash 8 times. Wash the resin with 20 mL of 20 mM Tris–HCl, pH 8.0, as above.
8. Transfer the resin into a 2 mL Eppendorf tube and top up to 1 mL with 20 mM Tris–HCl, pH 8.0. Add 100 μL of 250 mM glutathione and incubate overnight at 4 °C on a rotating tube mixer.
9. Transfer the suspension into a mini spin column and centrifuge for 4 min at 1000 × *g*, 4 °C. Collect the eluate and determine the protein concentration, for example, using Bradford protein assay. Aliquot and store at −80 °C. Additionally, run SDS-PAGE gel to determine the purified protein quality, together with a sample of the resuspended resin after elution with glutathione and the samples from previous steps (*see* **Note 9**).
10. Determine the activity of the recombinant PP2C phosphatase using a kit for assaying serine/threonine phosphatase activity, according to the protocol provided by the manufacturer.

    Calculate specific activity (U), where one U is the amount of phosphatase, which releases 1 picomole of phosphate per minute.

3.1.3 *In Vitro* Phosphatase Assay

1. Prepare samples on ice, by pipetting into 1.5 mL tube 1 μg of SnRK2 kinase, a specific amount of GST-PP2C (increasing amounts in subsequent samples, for example, 5, 10, and 25 activity units [U]), or GST as negative control (in the maximum amount used), 1.25 μL of 500 mM $MgCl_2$, 5 μL of 100 mM Tris–HCl (pH 7.5) and ddH_2O to a final volume of 25 μL.

2. Incubate the samples in a water bath for 30 min at 30 °C. Stop the reactions by adding 6.25 μL of 5× SDS sample buffer and incubating for 2 min at 95 °C.
3. Cool samples and spin down briefly, proceed to Subheading 3.1.4. or store at −20 °C.

3.1.4 In-Gel Kinase Assay

The following protocol is based on the method described by Wang and Zhu [19], with minor modifications.

1. Prepare 1 mm-space mini gel cast cassette and carefully clean glass plate with ethanol. Prepare the resolving gel mix (5 mL/gel) containing MBP, then cast the gel mix into the cassette (*see* **Note 12**) and overlay with a small amount of isopropanol to cut off air access. Leave the gel to polymerize for 40 min at RT. After that time gently remove the isopropanol with filtration paper avoiding disturbing the surface of the polymerized gel. Prepare the stacking gel mix (3 mL/gel), immediately pour it on top of the resolving gel and insert proper comb. Leave the gel to polymerize for 40 min at RT.
2. Load protein ladder marker (*see* **Note 13**) and 25 μL of samples from Subheading 3.1.3 on gel. Run the gel at 140 V at 4 °C until the 25 kDa marker runs off the gel (for approx. 2.5 h).
3. Prepare SDS removal solution (SRS, 300 mL per gel). Right before use supplement the SRS with DTT.
4. After the run take each gel from the cassette, place it in a plastic container with a lid and gently pour 100 mL of SRS. Shake the gels on an orbital platform with gentle motion (*see* **Note 14**) for 20 min, discard the SRS. Repeat this wash step 2 more times.
5. Prepare renaturation solution (300 mL per gel). Right before use supplement the renaturation solution with DTT. Carefully transfer the gel into a new container and gently pour 100 mL of renaturation solution. Shake the gels on orbital platform with gentle motion for 1 h at RT. Then replace the renaturation solution with a fresh portion and shake the gels on an orbital platform with gentle motion overnight (or at least 16 h) at 4 °C. After that time replace the renaturation solution with a fresh portion and incubate with gentle agitation for 1 h at RT.
6. Prepare kinase reaction solution (KRS, 100 mL per gel). Right before use supplement the KRS with DTT. Carefully replace renaturation solution with 70 mL of KRS and incubate gel for 30 min at RT with gentle agitation. After that time replace the KRS with 30 mL of a fresh portion and supplement it with 3.5 μL of [γ-^{32}P] ATP (35 μCi; *see* **Note 15**) and incubate for 5 min with gentle agitation. Then add 4.5 μL of 1 mM nonradioactive ATP (*see* **Note 16**) and incubate with gentle agitation for another 1.5 h at RT.

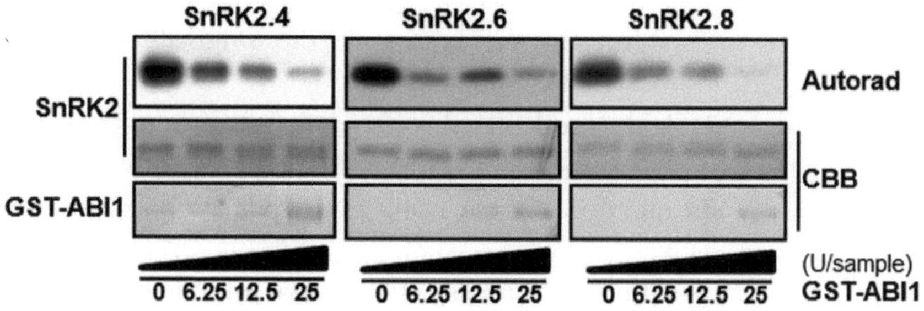

Fig. 1 ABI1 inhibits SnRK2 kinases in vitro. Recombinant SnRK2.4, SnRK2.6 (also known as SRK2E or OST1) or SnRK2.8 (also known as SRK2C) (1 μg) were incubated with increasing amounts of GST-ABI1 for 30 min at 30 °C and kinase activity was analyzed by in-gel kinase assay using MBP as substrate. Autorad, autoradiograph; CBB, Coomassie Brilliant Blue. (This figure was modified from Krzywińska E, Bucholc M, Kulik A, Ciesielski A, Lichocka M, Dębski J, Ludwików A, Dadlez M, Rodriguez PL, Dobrowolska G (2016) Phosphatase ABI1 and okadaic acid-sensitive phosphoprotein phosphatases inhibit salt stress-activated SnRK2.4 kinase. BMC Plant Biol. 16 (1):136 https:/creativecommons.org/licenses)

7. Prepare gel wash solution (GWS, 1 L per gel). Replace the KRS with 100 mL of GWS and incubate gel for 30 min at RT with gentle agitation. After first wash transfer the gel into a new container and repeat the washing 9 times (*see* **Note 17**).
8. Stain gel with Coomassie staining protocol and then wash them briefly (10–15 min) in 100 mL of preservation solution (*see* **Note 18**). Dry gels using gel drying film and detect radioactive signal using chosen method, for example, phosphorimager or autoradiographic film (*see* **Note 19**), (Fig. 1).

3.2 Immunocomplex Kinase Activity Assay for Determining the Effect of PP2C on SnRK2 Activity In Vivo

3.2.1 Preparation of Protein Extracts from Plant Material

1. Grind plant material to a powder in liquid nitrogen, for example, by using a mortar and pestle.
2. Transfer approximately 1 mL of ground material into a 2 mL tube (*see* **Note 20**) and add 300 μL of immunoprecipitation buffer supplemented with inhibitors, and let thaw on ice, vortexing from time to time.
3. Sonicate the samples 4 times for 30 s with a 30 s interval.
4. Centrifuge 20,000 × *g*, 4 °C for 20 min.
5. Determine the protein concentration, for example, using Bradford protein assay.

3.2.2 Immunocomplex Kinase Activity Assay

1. For each sample 25 μL of protein A-agarose slurry and 25 μL of the anti-SnRK2.4/SnRK2.10 antibodies are needed (*see* **Note 21**). Calculate the total volume for all the samples by multiplying by the number of samples plus adding 2 additional, to adjust for pipetting errors. Vortex the protein A-agarose (*see* **Note 6**) briefly and pipet the appropriate volume into a 1.5 mL tube. Spin down briefly using a benchtop mini centrifuge. Aspire the supernatant, add 1 mL of immunoprecipitation buffer to the agarose, vortex briefly and spin down as before.

Repeat this wash step 2 times. Resuspend the agarose in immunoprecipitation buffer, at the original volume before the washes (*see* **Note 22**).

2. Add the calculated volume of antibodies to the tube and incubate for 2 h at 4 °C on a rotating tube mixer.
3. Spin down briefly using a mini benchtop centrifuge.
4. Aspire the supernatant and wash the protein A-agarose-bound antibodies 5 times with 1 mL of immunoprecipitation buffer.
5. Resuspend the protein A-agarose-bound antibodies in immunoprecipitation buffer, at the volume before washes from **step 3** (*see* **Note 23**).
6. Pipet 500 μg of total protein (*see* Subheading 3.2.1) to a 1.5 mL tube for each protein sample, top up to 500 μL with immunoprecipitation buffer.
7. Add 50 μL of suspension of protein A-agarose-bound antibodies (*see* **step 5**), vortex briefly and incubate for 4 h at 4 °C on a rotating tube mixer.
8. Spin down briefly using a mini benchtop centrifuge.
9. Aspire the supernatant and resuspend pellet in 1 mL of immunoprecipitation buffer by vortexing. Repeat this wash step 3 more times.
10. Resuspend the pellet in 30 μL of immunoprecipitation buffer and add 5× SDS sample buffer.
11. Incubate at 95 °C for 2 min, let cool and spin down briefly.
12. Analyze the activity of immunoprecipitated kinases with in-gel kinase assay, as in Subheading 3.1.4 (Fig. 2) (*see* **Note 24**).

4 Notes

1. In this protocol the PP2C is purified as a GST-fusion protein. The epitope can also be cleaved off after binding of the protein to the resin, by thrombin or another selected protease. In that case select an appropriate pGEX vector for cloning, remembering to check if the recognition sequence of the chosen protease is not also present in the PP2C protein.
2. DTT should be added into the buffer freshly before use.
3. The use of Viscolase can be omitted, however its addition to resuspension buffer helps to decrease the lysate viscosity by degrading bacterial DNA and RNA.
4. The buffer can be prepared earlier and frozen at −20 °C, however in that case prepare it without DTT and PMSF and add them from stock solutions after thawing the buffer, freshly before use.

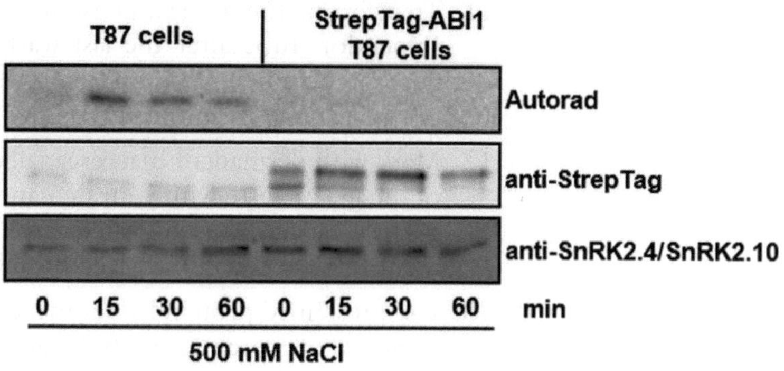

Fig. 2 ABI1 inhibits the activity of SnRK2 kinases in cells. Arabidopsis T87 cells and T87 cells expressing StrepTag-ABI1 were exposed to 500 mM NaCl. Activity of SnRK2.4/SnRK2.10 in the cell extracts was monitored by immunocomplex kinase activity assay, using MBP as substrate. The level of StrepTag-ABI1 and SnRK2.4/SnRK2.10 proteins was monitored by immunoblotting. Autorad, autoradiograph; CBB, Coomassie Brilliant Blue. (This figure was modified from Krzywińska E, Bucholc M, Kulik A, Ciesielski A, Lichocka M, Dębski J, Ludwików A, Dadlez M, Rodriguez PL, Dobrowolska G (2016) Phosphatase ABI1 and okadaic acid-sensitive phosphoprotein phosphatases inhibit salt stress-activated SnRK2.4 kinase. BMC Plant Biol. 16 (1):136 https:/ creativecommons.org/licenses/)

5. The immunocomplex kinase activity assay can be performed using custom-made antibodies raised against SnRK2, as in this protocol, or utilize available anti-SnRK2 antibodies (currently available from Agrisera). The protocol can also be used to analyze the activity of epitope-tagged kinases from plant material, in that case use specific antibodies recognizing the epitope.
6. Choose protein-A, protein-G, or protein-A/G agarose, depending on the species in which the antibodies used for immunoprecipitation where produced in.
7. At this point the bacterial pellet can be frozen and kept in −20 °C up to 1 month prior the purification.
8. Do not sonicate for longer periods of time, prolonged sonication can decrease the efficiency of the GST-fusion protein to bind to the resin.
9. Collecting the samples from all protein purification steps (resuspended bacteria, sonicated lysate, unbound lysate and purified protein eluate) and running them on SDS-PAGE gel helps to determine the quality of the purified protein and allows to optimize the proper steps of the protein purification process when needed.
10. Overall, 10 washes; collect a sample for SDS-PAGE analysis from the supernatant after the first wash only, discard the supernatant after the remaining washes.

11. Transferring the resin suspension into a 2 mL round-bottomed Eppendorf tube after the last washing step allows for better fluid circulation and thus allows thrombin to effectively cleave off the protein from the tag.
12. Mark with permanent marker small dot on the outside surface of the gel cassette with your demanded height of the resolving gel, it is especially useful when casting few gels for same experiment to have equally resolved proteins on all of them.
13. Use prestained protein ladder marker, which will enable seeing what size of proteins run off the gel and stopping the run at the appropriate time.
14. Remember to set the shaker to gentle settings, because any small disruptions and breaks in gel might be enlarged during too intensive agitation and gel handling will be very difficult in further steps.
15. While working with the [γ-^{32}P] ATP use barrier tips to prevent the contamination of the pipette. All steps which include working with [γ-^{32}P] ATP and [γ-^{32}P] ATP-contaminated solid materials and solutions are supposed to be performed in a properly adjusted working place shielded with special covers, according to local procedures regarding work with radioactively labeled materials and disposal of such materials. Please follow them carefully.
16. Nonradioactive ATP is added to reduce unspecific signal and background.
17. If you cannot perform all 10 washes during one working day you can leave the gel in fresh GWS overnight and continue the washing protocol on the next day.
18. This step is optional although our laboratory practice showed that short wash in preservation solution prevents gel breakage during drying. Just follow thoroughly the time indication since longer gel exposure to this solution ends in unwanted gel shrinkage.
19. Whether you will pick phosphorimager or autoradiographic film as a method of radioactive signal detection you should be aware you need to be supplemented with specific screen and or cassette for each detection method. In autoradiography it is not obligatory, but if signal from the samples is low phosphoscreen helps increase it.
20. To avoid thawing of the ground material while transferring to a tube, first place the tube in a tube holder partially immersed in liquid nitrogen and cool a laboratory spoon in liquid nitrogen before transferring the powder.
21. The volume of the antibodies used per sample depends on their reactivity, for example, a smaller amount will be used for affinity purified antibodies. Refer to the datasheet of the antibodies, as

a guideline use 5–25 μL of polyclonal antiserum or 5 μg affinity purified polyclonal antibody per 500 μg of protein sample. Dilute the antibody to 25 μL per sample to proceed as in the protocol, or adjust protocol volumes.

22. For example, pipet 250 μL of protein-A agarose into the tube, mark the level of the liquid with a permanent marker on the side of the tube, perform washes and resuspend the agarose in 250 μL of immunoprecipitation buffer (up to the mark).
23. For example, if 250 μL of protein-A agarose was preequilibrated and then 250 μL of anti-SnRK2 antibodies were added, mark the total level on the tube, perform washes and resuspend the agarose–protein complexes in 500 μL of immunoprecipitation buffer (up to the mark).
24. When comparing activity of kinases in extracts from, for example, wild-type plants (cells) and those with an altered level of PP2C, it is best to evaluate if the protein level of the studied SnRK2 is comparable in the samples, by Western blot.

Acknowledgments

This work was supported by National Science Centre (grant 2011/03/B/NZ3/00297).

References

1. Fuchs S, Grill E, Meskiene I, Schweighofer A (2013) Type 2C protein phosphatases in plants. FEBS J 280(2):681–693
2. Schweighofer A, Hirt H, Meskiene I (2004) Plant PP2C phosphatases: emerging functions in stress signaling. Trends Plant Sci 9(5): 236–243
3. Bhaskara GB, Wong MM, Verslues PE (2019) The flip side of phospho-signalling: regulation of protein dephosphorylation and the protein phosphatase 2Cs. Plant Cell Environ 42(10): 2913–2930
4. Hsu PK, Dubeaux G, Takahashi Y, Schroeder JI (2021) Signaling mechanisms in abscisic acid-mediated stomatal closure. Plant J 105(2):307–321
5. Chen K, Li GJ, Bressan RA, Song CP, Zhu JK, Zhao Y (2020) Abscisic acid dynamics, signaling, and functions in plants. J Integr Plant Biol 62(1):25–54
6. Yoshida T, Christmann A, Yamaguchi-Shinozaki K, Grill E, Fernie AR (2019) Revisiting the basal role of ABA - roles outside of stress. Trends Plant Sci 24(7):625–635
7. Mao X, Li Y, Rehman SU, Miao L, Zhang Y, Chen X, Yu C, Wang J, Li C, Jing R (2020) The sucrose non-fermenting 1-related protein kinase 2 (SnRK2) genes are multifaceted players in plant growth, development and response to environmental stimuli. Plant Cell Physiol 61(2):225–242
8. Fàbregas N, Yoshida T, Fernie AR (2020) Role of Raf-like kinases in SnRK2 activation and osmotic stress response in plants. Nat Commun 11(1):6184
9. Ma Y, Szostkiewicz I, Korte A, Moes D, Yang Y, Christmann A, Grill E (2009) Regulators of PP2C phosphatase activity function as abscisic acid sensors. Science 324(5930): 1064–1068
10. Park SY, Fung P, Nishimura N, Jensen DR, Fujii H, Zhao Y, Lumba S, Santiago J, Rodrigues A, Chow TF, Alfred SE, Bonetta D, Finkelstein R, Provart NJ, Desveaux D, Rodriguez PL, McCourt P, Zhu JK, Schroeder JI, Volkman BF, Cutler SR (2009) Abscisic acid inhibits type 2C protein phosphatases via the PYR/PYL family of

START proteins. Science 324(5930): 1068–1071

11. Fujii H, Chinnusamy V, Rodrigues A, Rubio S, Antoni R, Park SY, Cutler SR, Sheen J, Rodriguez PL, Zhu JK (2009) In vitro reconstitution of an abscisic acid signalling pathway. Nature 462(7273):660–664
12. Cutler SR, Rodriguez PL, Finkelstein RR, Abrams SR (2010) Abscisic acid: emergence of a core signaling network. Annu Rev Plant Biol 61:651–679
13. Vlad F, Rubio S, Rodrigues A, Sirichandra C, Belin C, Robert N, Leung J, Rodriguez PL, Laurière C, Merlot S (2009) Protein phosphatases 2C regulate the activation of the Snf1-related kinase OST1 by abscisic acid in Arabidopsis. Plant Cell 10:3170–3184
14. Umezawa T, Sugiyama N, Mizoguchi M, Hayashi S, Myouga F, Yamaguchi-Shinozaki K, Ishihama Y, Hirayama T, Shinozaki K (2009) Type 2C protein phosphatases directly regulate abscisic acid-activated protein kinases in Arabidopsis. Proc Natl Acad Sci U S A 106(41): 17588–17593
15. Soon FF, Ng LM, Zhou XE, West GM, Kovach A, Tan MH, Suino-Powell KM, He Y, Xu Y, Chalmers MJ, Brunzelle JS, Zhang H, Yang H, Jiang H, Li J, Yong EL, Cutler S, Zhu JK, Griffin PR, Melcher K, Xu HE (2012) Molecular mimicry regulates ABA signaling by SnRK2 kinases and PP2C phosphatases. Science 335(6064):85–88
16. Krzywińska E, Bucholc M, Kulik A, Ciesielski A, Lichocka M, Dębski J, Ludwików A, Dadlez M, Rodriguez PL, Dobrowolska G (2016) Phosphatase ABI1 and okadaic acid-sensitive phosphoprotein phosphatases inhibit salt stress-activated SnRK2.4 kinase. BMC Plant Biol 16(1):136
17. Krzywińska E, Kulik A, Bucholc M, Fernandez MA, Rodriguez PL, Dobrowolska G (2016) Protein phosphatase type 2C PP2CA together with ABI1 inhibits SnRK2.4 activity and regulates plant responses to salinity. Plant Signal Behav 12:e1253647
18. Frangioni JV, Neel BG (1993) Solubilization and purification of enzymatically active glutathione S-transferase (pGEX) fusion proteins. Anal Biochem 210(1):179–187
19. Wang P, Zhu JK (2016) Assessing kinase activity in plants with in-gel kinase assays. Methods Mol Biol 1363:189–197

Chapter 3

Reconstitution of the Core ABA Signaling in Protoplasts: Transcriptional Activators

Myung Ki Min, Rigyeong Kim, and Beom-Gi Kim

Abstract

Transient expression of genes in protoplasts has been used widely for purposes ranging from subcellular localization to promoter activity analyses. Here, we describe methods for reconstituting the abscisic acid (ABA) signaling pathway using a transient expression system in rice protoplasts. ABA signaling is monitored via reporter systems consisting of synthetic promoters and luciferase. Thus, the effects of each signaling component as well as complexes involved in ABA signaling can be characterized in rice protoplasts, overcoming many of the limitations that hamper efforts to identify biological functions of effector genes in whole plants. This protoplast-based transient assay system for ABA signaling thus provides valuable tools and knowledge for understanding complicated ABA signaling networks.

Key words ABA signaling, Rice protoplast, Transient expression, Signaling reporters, Signaling reconstitution

1 Introduction

Abscisic acid (ABA) is a major stress hormone in plants. It plays critical roles in plant adaptation to environmental stresses imposed by drought, salinity, low temperature, and radiation through physical and developmental processes such as seed dormancy, seed germination, growth retardation, and closure of stomata [1]. The ABA signaling pathway consists of four major components in terms of transcriptional regulation. Firstly, transcription factors containing a basic leucine zipper domain (bZIP) are mainly involved in the induction of ABA-responsive gene expression. Secondly, SNF1-related serine/threonine protein kinases 2 s (SnRK2s) play as a positive regulator for several ABA responses such as activation of the bZIP transcription factors and stomatal closure. Functional orthologs of SnRK2s in rice are osmotic stress/ABA-activated protein kinases (SAPKs). Thirdly, clade A of type 2C protein phosphatases (PP2Cs) are well known as a negative regulator in ABA

Takuya Yoshida (ed.), *Abscisic Acid: Methods and Protocols*,
Methods in Molecular Biology, vol. 2462, https://doi.org/10.1007/978-1-0716-2156-1_3,

signaling pathway. Those dephosphorylate SAPKs resulting in inhibition of ABA-dependent signal pathway. Fourthly, ABA receptors named as PYRABACTIN RESISTANCE1/PYRABACTIN RESIS TANCE LIKE/ REGULATORY C OMPONENTS OF ABA RECEPTOR (PYR1/PYL/RCAR) bind ABA together with PP2Cs to form an ABA receptor complex and start to transduce the signaling when ABA concentration goes up in plant cell [2, 3]. Each component has several homologs with different activities and functions. The functions of each gene have been identified using knockout mutants or in vitro activity assays in rice [4]. However, their redundancies make full characterization difficult. Transient assays using protoplasts offer the advantages of examining various signaling pathway components together and quickly while using an in vivo cell system. Reporters using promoters that respond precisely to ABA are efficient tools for monitoring the effects of these components. Here, we constructed an ABA-responsive luciferase reporter in rice using the promoter region (1.6 kb) of *OsRab16A* (LOC_Os11g26790), which is highly induced by ABA [5]. This reporter is highly responsive to transcriptional activators involved in ABA signaling such as OsbZIPs, and we used it to identify characteristics of these effectors through reconstitution of ABA signaling components. We then constructed a second version of the reporter that is more sensitive to ABA by using a synthetic promoter consisting of 2× -dehydration-responsive element (DRE) and ABA-responsive element (ABRE) derived from *OsRab16A* [6].

We successfully reconstituted the ABA signaling components using a transient gene expression system in rice protoplasts. The system has three main components. The first is the sensitive ABA reporter, the second is an efficient polyethylene glycol (PEG)-mediated protoplast transformation system, and the third is a vector suitable for cloning signaling effectors for transient expression in rice protoplasts. We describe detailed protocols for utilizing these components and provide examples for reconstituting the ABA signaling components in rice protoplasts. This high-throughput, rapid analysis tool is powerful for screening signaling components *in planta* and hypothesizing their functions and networks in plant cells.

2 Materials

2.1 Vector Constructs and Preparing of Plasmid DNA

1. Plasmid purification kits (e.g., Qiagen or equivalent) (*see* **Note 1**).
2. ABA-responsive reporter: The ABA-responsive firefly luciferase (fLUC) reporter construct pABRE-DRE::fLUC was constructed by fusing an ABA-responsive synthetic promoter and fLUC (*see* **Note 2**) (Fig. 1).

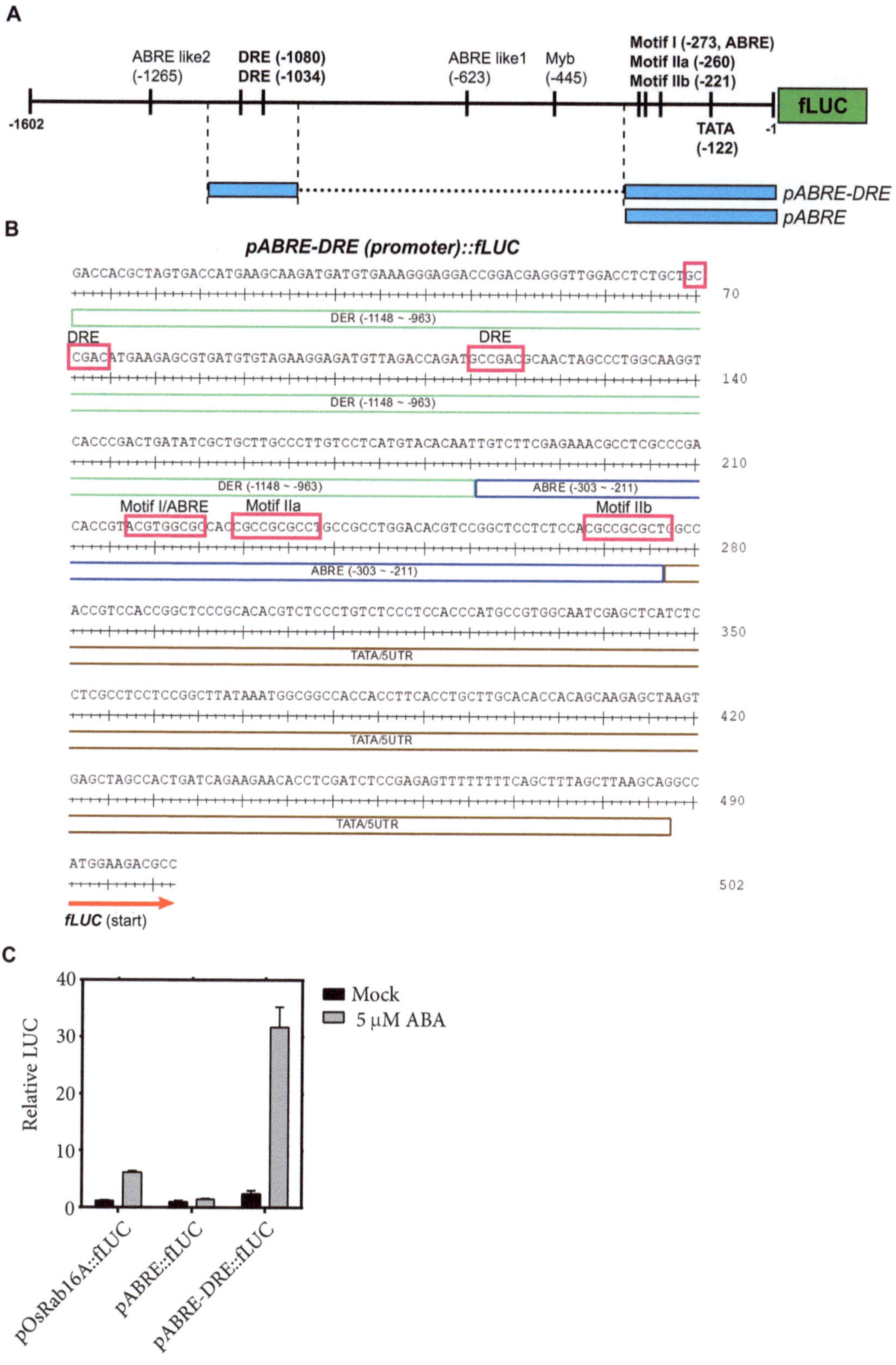

Fig. 1 Generation of pABRE-DRE::fLUC. (**a**) Scheme showing ABRE and DRE components in the *OsRab16A* promoter. (**b**) Sequence of pABRE-DRE with ABRE and DRE *cis*-elements and chimeric sequences indicated. (**c**) Response of the reporter plasmids (*pABRE-DRE::fLUC* and *pABRE::fLUC*) to ABA; error bars are SD. Luciferase activity is given relative to that in control samples

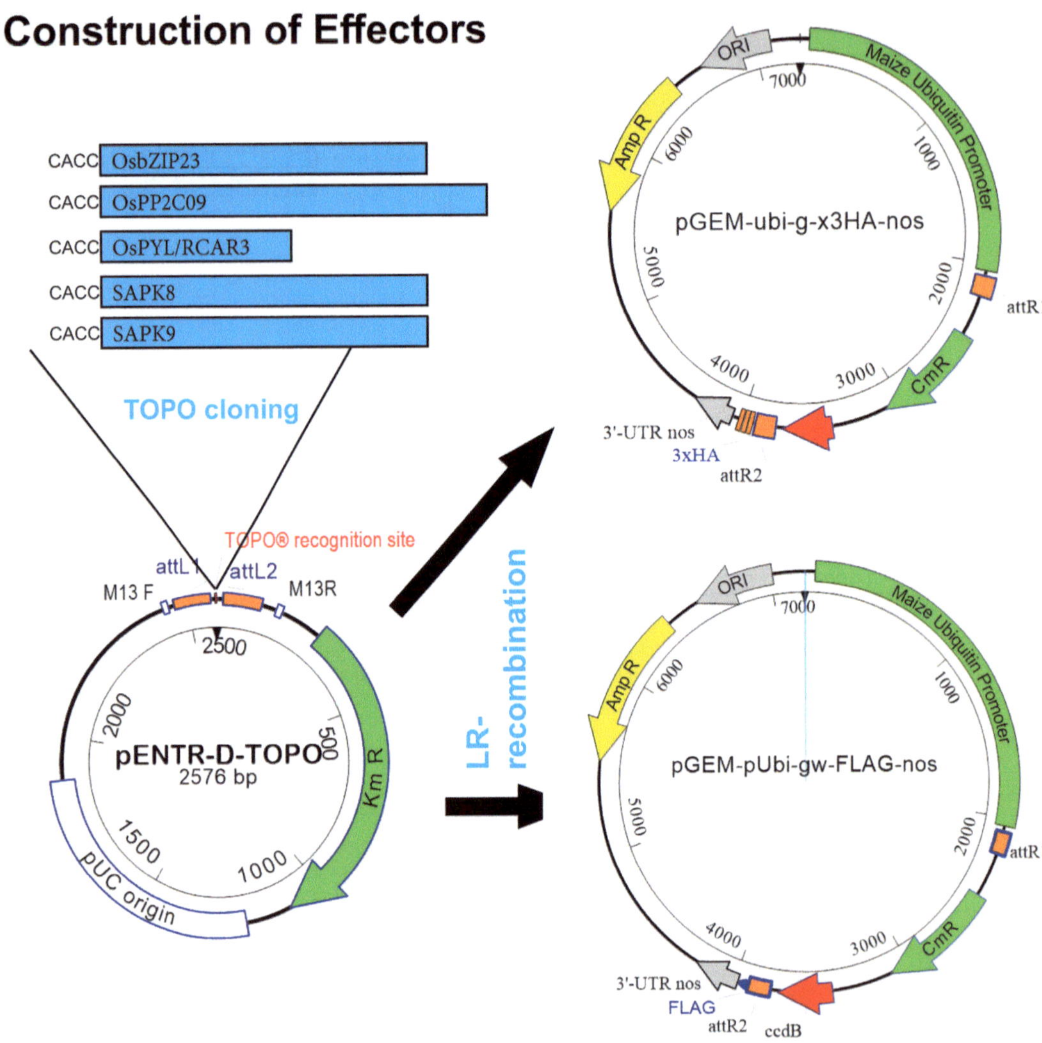

Fig. 2 Construction of effector plasmids and vectors used for transient expression in rice protoplasts

3. Vectors for transient expression in rice protoplasts: Plasmids used for overexpression of effector genes in rice protoplasts were constructed using the gateway system (Thermo Fisher Scientific), the maize ubiquitin promoter, and 3xHA or FLAG tag (Fig. 2).
4. Effector plasmid constructs: *OsbZIP23* (LOC_Os02g52780), *OsPP2C09* (LOC_Os01g62760), *SAPK8* (LOC_Os03g55600), *SAPK9* (LOC_Os12g39630), and *OsPYL/RCAR3* (LOC_Os02g15640) CDS were used to clone into the overexpression vector in this manuscript (*see* **Note 3**).
5. Reporter plasmids mixture: *pABRE-DRE::fLUC* and *pUbi::rLUC* in a ratio of 10:1.

2.2 Preparation of Plants

1. Plant cultivar: *Oryza sativa* cv. Dongjin.
2. 2% (w/v) sodium hypochlorite.
3. 70% (v/v) ethanol.
4. Distilled water (DW).
5. Growth medium: 2.15 g Murashige and Skoog (MS) medium without sucrose and 0.5 g 2-(N-morpholino)ethanesulfonic acid (MES), adjusted to pH 5.7 using KOH, for 1 L culture medium (1/2MS). Add 0.4% (w/v) Phytagel before autoclaving at 121 °C for 15 min to solidify medium.
6. Tissue culture vessels, 72 × 72 × 100 mm.
7. Rice huller.

2.3 PEG-Mediated Transformation of rice Protoplasts

1. Enzyme solution for rice: 1.5% cellulose R-10 (e.g., Yakult Pharmaceutical INC. or equivalent), 0.75% macerozyme R-10 (e.g., Yakult Pharmaceutical INC. or equivalent), 0.6 M mannitol, and 10 mM MES, adjusted to pH 5.7 using KOH. Sterilize using a 0.45 μm syringe filter. The enzyme solution can be stored frozen; reactivate at 60 °C for 20 min before using and then cool slowly at room temperature. Prepare finally enzyme solution for use by adding 0.1% bovine serum albumin (BSA), 3.4 mM $CaCl_2$, 5 μM β-mercaptoethanol, and 50 μg/mL ampicillin (*see* **Note 4**).
2. W5 solution: 154 mM NaCl, 125 mM $CaCl_2$, 5 mM KCl, 5 mM glucose, and 3 mM MES, adjusted to pH 5.7 using KOH. Autoclave (121 °C, 15 min) and store at room temperature.
3. 23% sucrose solution: Dissolve 23% (w/v) sucrose in W5 solution or distilled water.
4. MaMg solution: 600 mM Mannitol, 15 mM $MgCl_2$, and 5 mM MES, adjusted to pH 5.7 using KOH. Use after autoclaving at 121 °C for 15 min and store at room temperature.
5. PEG solution: 400 mM mannitol, 100 mM $Ca(NO_3)_2$, 40% PEG 6000 (*see* **Note 5**). Make fresh for each experiment.
6. 5 mM ABA in DMSO (*see* **Note 6**).
7. 14 mL round-bottom polypropylene tube.
8. Hemocytometer.
9. Swinging-rotor centrifuge.
10. Cell dissociation sieve.
11. Shaker.
12. Sharp blade.
13. Rotatory drum mixer.

2.4 Detection of LUC Activity

1. Dual-reporter assay reagents (e.g., Promega Dual-Luciferase® Reporter Assay System or equivalent).
2. Microplate Luminometer (e.g., Promega GloMax 96 or equivalent).
3. 96 well plate.

2.5 Detection of Effector Proteins

1. 5× sample buffer: 25% glycerol, 2% SDS, 200 mM DTT, 0.1% bromophenol blue, 60 mM Tris–HCl (pH 6.8).
2. Instruments and consumables for SDS phase gel electrophoresis.
3. Instruments and consumables for Western blot analysis.
4. Antibodies: monoclonal anti-flag M2 peroxidase mouse antibody (Sigma), monoclonal anti-HA rat antibody (clone 3F-10, Roche).

3 Methods

3.1 Plant Growth

1. Hull rice seeds using a rice huller (Fig. 3a). Sterilize hulled seeds with 70% (v/v) ethanol for 30 s and 2% sodium hypochlorite for 40 min, sequentially. Wash seeds 5–6 times with sterilized DW.
2. Plant sterilized seeds on 1/2MS medium without sucrose and containing 0.4% Phytagel in vessels. Incubate at 28 °C under dark conditions for 8 days and then transfer to long-day conditions (16/8 h light, approximately 41 μmol m^{-2} s^{-1}/dark) at 28 °C for 2 days (Fig. 3b).

3.2 PEG-Mediated Transformation of Rice Protoplasts

1. Chop a bundle of rice seedlings (about 25–35 seedlings) into pieces of around 1 mm using a sharp blade (*see* **Note** 7 and Fig. 3c).
2. Quickly transfer the chopped seedling pieces to freshly prepared enzyme solution. Incubate the soaked seedling pieces at 28 °C for 4 h under dark conditions with gentle rocking (15° tilt angle and 20–30 rpm) (*see* **Note 4** and Fig. 3d, e).
3. Filter the enzyme solution containing protoplasts through a cell dissociation sieve to remove tissue debris (Fig. 3f). Dilute with 3–4 volumes of W5 solution and mix well. Precipitate protoplasts by swinging-rotor centrifugation (100 × *g*, 15 min).
4. Remove supernatant and then resuspend protoplasts with 10 mL of W5 solution. Carefully place W5 solution containing protoplasts on top of the same volume of 23% (w/v) sucrose solution (Fig. 3g) in a 14 mL round-bottom tube. After centrifugation (100 × *g*, 10 min, 28 °C), transfer carefully the

Fig. 3 Photographic diagram for PEG-mediated transient expression assay. (**a**) Rice hullerswipper. (**b**) Rice seedlings. (**c**) Preparation of samples for enzyme treatment. (**d**) Rice strips submerged in enzyme solution. (**e**) Incubation on a twister for isolation of protoplasts. (**f**) Tissue debris removal using a cell dissociation sieve. (**g**, **h**) Removal of broken protoplasts (in sucrose layer after centrifugation) using a sucrose gradient. (**i**, **j**) Use of a rotatory drum mixer for mixing DNA and protoplasts in MaMg and PEG solution. (**k**) Incubation of protoplasts in W5 solution at 28 °C. (**l**) Protoplasts in PLB are aliquoted into a 96-well plate. (**m**) GloMax 96 Microplate Luminometer

protoplast layer to a new tube (*see* **Note 8** and Fig. 3h). Resuspend protoplasts with around 10 mL of W5 solution for washing.

5. Centrifuge protoplasts at 100 × *g* and 28 °C for 5 min and remove supernatant. Resuspend protoplasts with MaMg solution to 4–5 × 10^6 cells/mL. Count the number of protoplasts using a hemocytometer.

6. Aliquot reporter plasmid mixture (5 μg) into 14 mL round-bottom tubes. Add effector plasmids as indicated in Fig. 4. Amounts of effector DNA must be optimized according to

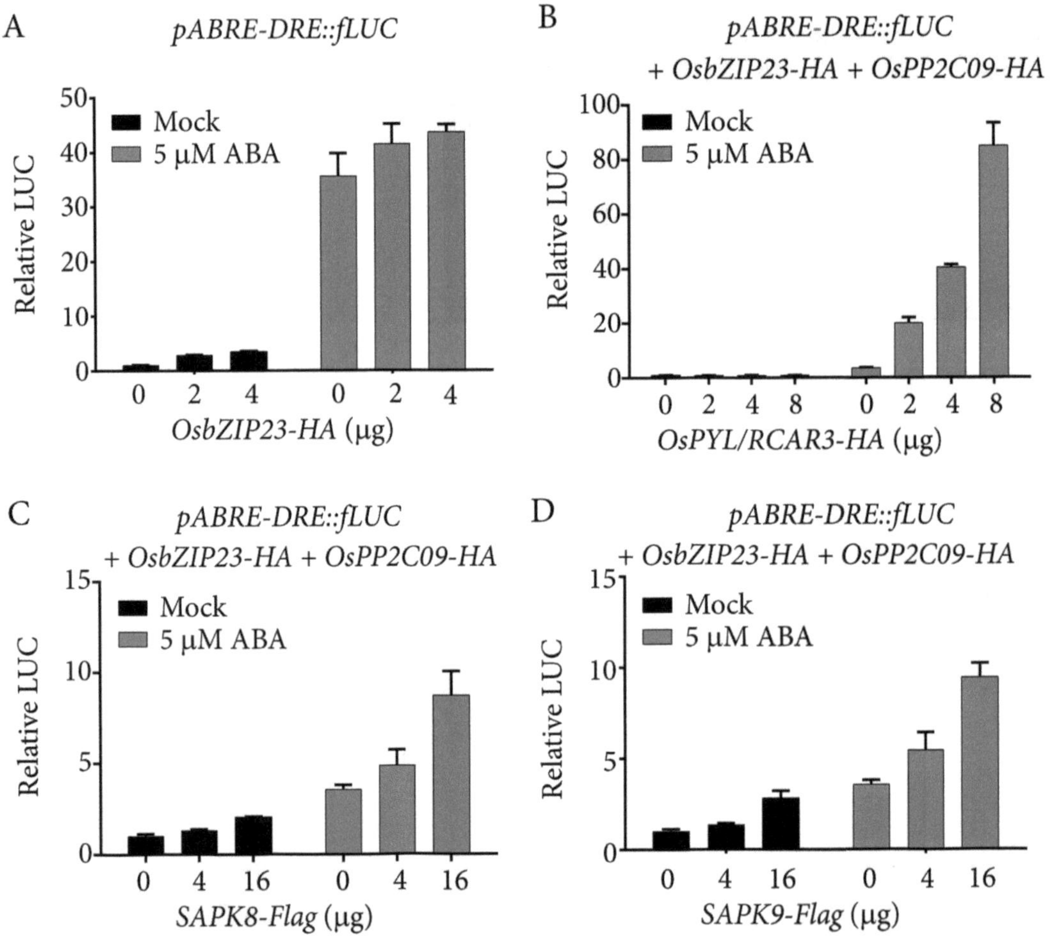

Fig. 4 ABA signaling reconstruction using several ABA signaling components. (**a**) LUC reporter activity induced by overexpression of a rice bZIP transcription factor. (**b**) ABA receptor (OsPYL/RCAR3) activates the pABRE-DRE promoter only under ABA conditions. (**c**, **d**) SAPK8 and SAPK9 are positive regulator for ABA signaling. Error bars are SD. These experiments were replicated three independent times

experimental purposes. Carefully add **300** μL of protoplasts resuspended in MaMg solution into the tubes and mix the DNA carefully (*see* **Note 9**).

7. Add 330 μL of PEG solution to the DNA/protoplast mixture (Fig. 3i) and mix carefully until the gradient disappears (*see* **Note 10**).
8. After incubation for 20 min at 28 °C, add an equal volume of W5 solution (0.63 mL), dispensing carefully along the tube surface to reduce protoplast disruption. Mix gently until the gradient disappears. Allow 10 min from the first addition of W5 solution before adding an equal volume of W5 solution (1.2 mL). And then after 10 min, add an equal volume of W5 solution (2.5 mL) again (*see* **Note 10** and Fig. 3J).

9. Mix for 10 min and then incubate for 30 min at 28 °C. Centrifuge the protoplasts at 100 × *g* and 28 °C for 5 min using a swing-rotor centrifuge. Remove the supernatant and resuspend the protoplasts with 3 mL of W5 solution (*see* **Note 11** and Fig. 3K).
10. Divide each protoplast sample into two or three to confirm protein expression: a mock sample, one treated with ABA (5 μM) and additional sample for western blot analysis (*see* **Note 6**). Incubate at 28 °C under dim light (approximately 3 μmol m^{-2} s^{-1}) conditions for 15–22 h.

3.3 Detection and Analysis of LUC Activity

Here, we describe an example procedure using some commercially available dual-reporter assay reagents, and please modify and adjust the following steps according to the manufacturer's instructions of the reagent and the luminometer.

1. Remove W5 solution of 1–1.5 mL of protoplasts by centrifugation at 3000 × *g* for 1–2 min. Briefly suspend with 50 μL of 1× Passive Lysis Buffer (PLB). Keep on ice at all times (*see* **Note 12**).
2. Immediately aliquot 10 μL of each sample into three wells of a 96-well plate (Fig. 3L).
3. Detect LUC activity using a microplate luminometer as described in the manufacturer's instructions (Fig. 3m). Measure luminescence at three time-points: before injection (blank value) to calibrate values; after injection of luciferase Assay Reagent II containing firefly luciferase substrate to check induced values by ABA signaling (firefly luciferase activity); after injection of Stop & Glo® Reagent containing Renilla luciferase substrate (Renilla luciferase activity) to check transformation efficiency. All detections performed for 5 s with 1 s interval after injections in each step.
4. Calculate LUC activity relative to control values using the following formula:

   [(2)–(1)]/[(3)–(1)]. The (1) is blank value, the (2) is firefly luciferase activity, and the (3) is Renilla luciferase activity.
5. The data value is displayed relatively compared with control value (Fig. 4).

3.4 Detection of Effector Proteins

1. Centrifuge the remained 1 mL of protoplasts at 3000 × *g* for 1–2 min and then removed the W5 solution.
2. Dissolve the pellets in 5× sample buffer for SDS-PAGE.
3. Perform western blot analysis (*see* **Note 13**).

4 Notes

1. All plasmids must be purified to high quality to minimalize monomeric nucleotides and RNAs. We usually use commercially available purification kits (such as the Midi- or Maxi-DNA purification kit from Macherey-Nagel or Qiagen). However, DNA purification methods such as cesium gradient centrifugation for DNA separation or PEG-mediated plasmid purification are also suitable.
2. The *cis*-elements ABRE and DRE are from the promoter of *OsRab16A* and were assembled by recombining a 182-bp sequence including two repeats of 5′-GCCGAC-3′ (DRE) and a 100-bp sequence including 5′- ACGTGGCGC -3′ (ABRE). The pABRE fragment was amplified using *Bam*HI-pABRE-F and *Nco*I-pRab16A-R primers and then inserted into the fLUC vector using *Nco*I and *Bam*HI restriction sites. The pABRE-DRE fragment was generated by amplifying the DRE fragment (using *Bam*HI-pDRE-F/p(ABRE)-DRE-R primers) and the ABRE fragment (using p(DRE)-ABRE-F/ *Nco*I-pRab16A-R primers) and then conjugating the two fragments by PCR amplification using *Bam*HI-pDRE-F/ *Nco*I-pRab16A-R. The amplified fragment was inserted into the fLUC vector using *Bam*HI and *Nco*I restriction sites. *See* Table 1 for primer sequences. In tests, the reporter construct *pABRE::fLUC* containing only ABRE did not respond to ABA; however, the construct *pABRE-DRE::fLUC* containing DRE and ABRE showed a stronger response to ABA than the full length *pOsRab16A* (Fig. 1C) [6].
3. Several effectors for ABA signaling reconstitution have already been reported in rice and Arabidopsis [5, 7].

Table 1
Primers for plasmid construction

Name	Forward primer Reverse primer	Sequence
pABRE	*Bam*HI-pABRE-F	TCGGATCCTGTCTTCGAGAAACGCCTC
	*Nco*I-pRab16A-R	TGCCATGGCCTGCTTAAGCTAAAGCTGA
pABRE-DRE	p(DRE)-ABRE-F	(CCTCATGTACACAAT)TGTCTTCGAGAAACGCCTCG
	*Nco*I-pRab16A-R	TGCCATGGCCTGCTTAAGCTAAAGCTGAA
	*Bam*HI-pDRE-F	TCGGATCCGACCACGCTAGTGACCATGA
	p(ABRE)-DRE-R	(CGTTTCTCGAAGACA)ATTGTGTACATGAGGACAAGGG
	*Bam*HI-pDRE-F	TCGGATCCGACCACGCTAGTGACCATGA
	*Nco*I-pRab16A-R	TGCCATGGCCTGCTTAAGCTAAAGCTGAA

4. Ampicillin (50 μg/mL) is used to prevent bacterial growth in the enzyme solution and the W5 solution for the final incubation. $CaCl_2$ and $Ca(NO_3)_2$ can be prepared as 1 M stock solutions. Leaf sheaths of around 50 seedlings are used per 25 mL of enzyme solution.
5. Choose PEG with molecular mass of 4000 Da or more (6000 or 8000). Dissolve by microwaving for a short time or melt at 60 °C with a long incubation (over 3 h). Allow to cool sufficiently at room temperature before use. Dissolve mannitol for 1 M stock solution by heating (60 °C) or brief microwave treatment.
6. Dissolve ABA in DMSO. Do not dissolve ABA in solvents such as ethanol and methanol because those can shorten the life of protoplasts and invoke stress responses [8].
7. Use of a blunt blade leads to destruction of cells around where the blade passes. Our experience has confirmed that protoplast isolation efficiency decreases when the number of broken cells increases. We recommend using a sharp blade such as a surgical blade.
8. We recommend using a pipette tip with the end cut off to manipulate protoplasts.
9. We typically use effectors in the range 1–10 μg. However, PP2Cs suppress the signal very strongly and so are used in the 0.1–1 μg range. The number of protoplasts and size of tubes can be varied. However, we concluded that 300 μL of protoplasts and 14 mL round-bottom tubes worked best in our experiments. At this time, we do not recommend using more than 30 μg DNA.
10. We mixed the protoplasts and PEG solution using a rotatory drum mixer at 10 rpm and 15° tilt angle for 10 min at 28 °C (Fig. 3J). These mixing conditions are the same as in the dilution step (adding W5).
11. In our experience, short incubation times induce low transformation efficiency and long incubation times affect cell viability.
12. The components of the Promega Dual-Luciferase Assay System.
    (a) Passive lysis buffer (PLB) for suspension of protoplasts

    When use it, do not freeze or vortex the sample vigorously after PLB treatment because the luciferase activity will drop abruptly during the measurement. The PLB is a buffer for animal cells for storing and suspending the cells attached to the plate. If protoplasts break during sample preparation, the half-life of luciferase during the assay is very short.

(b) Luciferase Assay Reagent II containing firefly luciferase substrate.

(c) Stop & Glo® Reagent containing Renilla luciferase substrate

For inactivation of firefly luciferase and provides for Renilla luciferase.

13. To analyze characteristics of effectors in ABA signaling precisely, we need to confirm at least once that ABA signaling effector DNAs are transfected into protoplasts and expressed the effector proteins. We recommend detecting proteins using immunoblotting analysis because the efficiency of protein biosynthesis is quite diverse among effectors. We constructed HA or FLAG epitope tagging vectors for transient overexpression in rice protoplasts. All effector proteins have conjugated epitopes such as HA or FLAG. Protein accumulation and quantity can be confirmed through immunoblotting analysis using the remaining samples. Only 1 mL of sample is required for confirming luciferase activity. We do not recommend using all 3 mL for immunoblotting analysis. We have experienced good results performing immunoblotting analysis with 1–1.5 mL of sample.

We used monoclonal anti-flag M2 peroxidase mouse antibody for FLAG detection and monoclonal anti-HA rat antibody for HA detection. We confirmed that the amount of protein increased with the amount of DNA used (data not shown).

Acknowledgments

This work was supported by the Agenda program (PJ01483801) and New breeding technologies development program (PJ014773) of the Rural Development Administration.

References

1. Vishwakarma K, Upadhyay N, Kumar N, Yadav G, Singh J, Mishra RK, Kumar V, Verma R, Upadhyay R, Pandey M (2017) Abscisic acid signaling and abiotic stress tolerance in plants: a review on current knowledge and future prospects. Front Plant Sci 8:161
2. Park S-Y, Fung P, Nishimura N, Jensen DR, Fujii H, Zhao Y, Lumba S, Santiago J, Rodrigues A, Tsz-fung FC (2009) Abscisic acid inhibits type 2C protein phosphatases via the PYR/PYL family of START proteins. Science 324(5930):1068–1071
3. Ma Y, Szostkiewicz I, Korte A, Moes D, Yang Y, Christmann A, Grill E (2009) Regulators of PP2C phosphatase activity function as abscisic acid sensors. Science 324(5930):1064–1068
4. Tian X, Wang Z, Li X, Lv T, Liu H, Wang L, Niu H, Bu Q (2015) Characterization and functional analysis of pyrabactin resistance-like abscisic acid receptor family in rice. Rice 8(1):1–13
5. Kim N, Moon SJ, Min MK, Choi EH, Kim JA, Koh EY, Yoon I, Byun MO, Yoo SD, Kim BG (2015) Functional characterization and reconstitution of ABA signaling components using

transient gene expression in rice protoplasts. Front Plant Sci 6:614

6. Min MK, Kim R, Moon S-J, Lee Y, Han S, Lee S, Kim B-G (2020) Selection and functional identification of a synthetic partial ABA agonist, S7. Sci Rep 10(1):4
7. Fujii H, Chinnusamy V, Rodrigues A, Rubio S, Antoni R, Park S-Y, Cutler SR, Sheen J, Rodriguez PL, Zhu J-K (2009) In vitro reconstitution of an abscisic acid signalling pathway. Nature 462(7273):660–664
8. Hann CT, Bequette CJ, Dombrowski JE, Stratmann JW (2014) Methanol and ethanol modulate responses to danger- and microbe-associated molecular patterns. Front Plant Sci 5:550–550

Chapter 4

Affinity Purification of Ubiquitinated Proteins Using p62-Agarose to Assess Ubiquitination of Clade A PP2Cs

Jose Julian, Alberto Coego, Abdulwahed F. Alrefaei, and Pedro L. Rodriguez

Abstract

Certain E3 ubiquitin ligases play a key role in the abscisic acid (ABA) pathway by targeting clade A type 2C protein phosphatases (PP2Cs) for degradation. At early stages of ABA signaling, degradation of PP2Cs is a complementary step to PP2Cs inhibition by ABA receptors. At later steps, protein levels of PP2Cs are increased as a negative feedback mechanism. Subsequently, E3 ligases targeting PP2Cs are critical to recover the basal PP2C levels and reset the ABA signaling. BTB/POZ AND MATH DOMAIN proteins (BPMs) are substrate adaptors of a multimeric cullin3-RING based E3 ligase and target for degradation clade A PP2Cs. In this chapter, we provide a detailed protocol to assess the ubiquitination of PP2CA, a clade A PP2C, mediated by BPMs using agarose-immobilised p62-derived ubiquitin-associated (UBA) domain, which efficiently binds ubiquitinated proteins.

Key words ABA, PP2C, p62, BPM, E3 ubiquitin ligase, Pulldown, Affinity purification

1 Introduction

Enhanced ABA levels produced in response to drought are perceived by the PYRABACTIN RESISTANCE1 (PYR1)/PYR1-LIKE (PYL)/REGULATORY COMPONENTS OF ABA RECEP TORS(RCAR) family of ABA receptors and clade A PP2Cs such as HYPERSENSITIVE TO ABA 1 (HAB1), ABA INSENSITIVE 1 (ABI1) and PP2CA, also known as ABA HYPERSENSITIVE G ERMINATION 3 (AHG3) [1, 2]. PYR/PYL/RCAR ABA receptors inhibit competitively clade A PP2Cs to activate ABA signaling [3–5]. Additional mechanisms to relief PP2C repression involve degradation of PP2Cs via the 26S proteasome [6–8]. For instance, ubiquitination and degradation of ABI1 is mediated by PLANT U-BOX 12 (PUB12) and PUB13 E3 ligases [6]. This ubiquitination is dependent on ABA receptors, and is strictly dependent on

Takuya Yoshida (ed.), *Abscisic Acid: Methods and Protocols*,
Methods in Molecular Biology, vol. 2462, https://doi.org/10.1007/978-1-0716-2156-1_4,

ABA for dimeric receptors and enhanced by ABA for monomeric receptors. PUB12/13 are unable to interact with other clade A PP2Cs, which suggested that other E3 ligases would be required to target the rest of clade A PP2Cs. For instance, RING DOMAIN LIGASE 1 (RGLG1) and RGLG5 RING-type E3 ligases mediate PP2CA, HAB2 and ABI2 ubiquitination, which is enhanced in vivo by the presence of ABA and ABA receptors [7, 9]. In addition to RGLG1 and RGLG5, we also identified BTB/POZ AND MATH DOMAIN proteins (BPMs) as PP2CA-interacting proteins, including as well major clade A PP2Cs such as HAB1 and ABI1 [8]. BPMs are substrate adaptors of the multimeric cullin3 (CUL3)-RING-based E3 ligases (CRL3s) and constitute a six-member gene family in Arabidopsis, BPM1 to 6 [10].

The regulation of protein stability of clade A PP2Cs has high importance in the regulation of ABA signaling because these proteins are key negative regulators of the pathway [11]. For the sake of clarity, we can distinguish several steps in ABA signaling (Fig. 1): (a) the relief of PP2C inhibition and activation of subfamily III SNF1-related protein kinase 2 s (SnRK2s); (b) ABA response at plasma membrane and nucleus through SnRK2-dependent phosphorylation of different targets, which leads to a negative feedback mechanism that involves the accumulation of newly synthesized PP2Cs; and finally, (c) the resetting of basal PP2C protein levels (PP2C proteostasis) [8]. E3 ubiquitin ligases play a major role for dynamic regulation of PP2C protein levels, which contributes to the activation and resetting of ABA signaling. The ubiquitination of clade A PP2Cs enhanced by ABA and ABA receptors helps in the activation of ABA response. Subsequently, when PP2Cs accumulate to attenuate ABA signaling, E3 ligases are critical to return clade A PP2Cs to their basal protein levels under nonstress conditions.

P62, also called SEQUESTOSOME1 (SQSTM1), is a selective cargo receptor for autophagy and possesses several structural domains, one of which is an UBA domain that efficiently recognizes ubiquitin (Ub) [18, 19]. This UBA domain is useful for immunoprecipitation and affinity purification of multiubiquitinated proteins and mixtures of ubiquitin conjugates; for instance, in the analysis of ubiquitinated ABA receptors or PP2Cs [20]. In this chapter, we will describe the use of agarose-immobilized p62-derived UBA domain (comprising the p62 residues 387–436) to perform a pulldown of ubiquitinated proteins from a plant extract, followed by immunoblot analysis of PP2CA. This assay enables the assessment of PP2CA ubiquitination in plant protein extracts obtained from wild type or *bpm* mutants, namely in two mutant lines, the *bpm3 bpm5* double mutant and an *amiR-bpm* line. This *amiR-bpm* line is impaired in the expression of *BPM1*, *BPM4*, *BPM5*, and *BPM6*. Plants are previously treated with ABA (to promote the accumulation of newly synthesized PP2Cs) and MG-132 (to inhibit the degradation of ubiquitinated

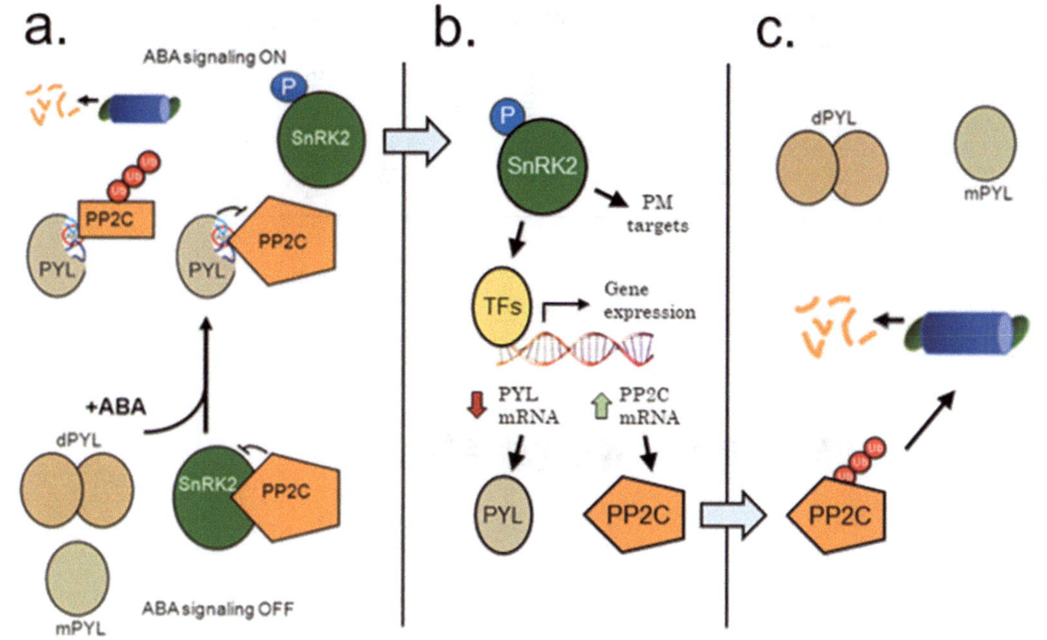

Fig. 1 Illustration of different steps in ABA signaling. (**a**) Activation of ABA signaling by formation of the receptor (PYL)-ABA-phosphatase (PP2C) complexes, which releases subfamily III SnRK2s from PP2C inhibition [12, 13]. Subfamily III SnRK2s are then phosphorylated, which requires RAF-like MAPKKKs [14–16]. PP2Cs are also degraded during this step, which is enhanced by the formation of ternary complexes with ABA receptors. (**b**) ABA response is initiated in plasma membrane (PM) and nucleus through SnRK2-dependent phosphorylation of different targets, which includes phosphorylation and activation of ABRE binding factors (ABFs) (also known as ABRE-binding (AREB) proteins). These transcription factors (TFs) are crucial to activate ABA-responsive gene expression. Interestingly, ABFs also activate PP2C transcription [17], which leads to accumulation of newly synthesized PP2Cs as a negative feedback mechanism (together with downregulation of some ABA receptors). (**c**) Resetting of PP2C basal levels is achieved by activation of E3s that target PP2Cs for degradation, while dimeric (dPYL) and monomeric (mPYL) receptors would be ready for a new round of ABA perception

proteins). Interestingly, plants impaired in BPMs accumulate a higher amount of PP2CA, from which a reduced fraction is ubiquitinated compared to wild type (Fig. 2).

2 Materials

2.1 Plant Material and Growth Conditions

1. 70% Ethanol solution, supplemented with 0.1% Triton.
2. Absolute Ethanol.
3. MS medium (for 1 L): 4.33 g Murashige and Skoog Basal Medium, 10 g Sucrose, adjust pH to 5.7 with KOH, for solid plates add 8 g Phytoagar before autoclaving.
4. Plant growth chamber.

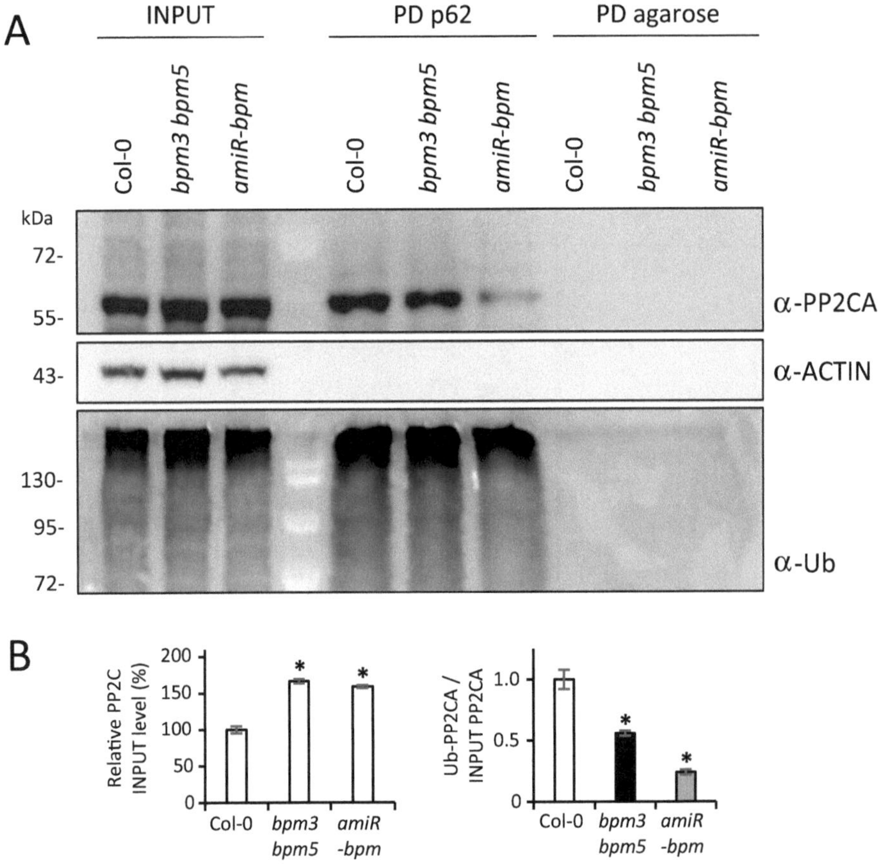

Fig. 2 In vivo ubiquitination of PP2CA is diminished in *bpm3 bpm5* and *amiR-bpm* mutants compared to Col-0. (**a**) Immunoblot analyses using α-PP2CA, α-ACTIN and α-Ub antibodies. α-PP2CA antibodies detect mostly nonubiquitinated and monoubiquitinated forms of PP2CA. ACTIN is used as a loading control of the INPUTs and α-Ub to confirm equivalent recovery of ubiquitinated proteins after pulldown (PD) with p62 agarose. The PD with empty agarose serves as a negative control. (**b**) Left histogram, PP2CA adjusted density signal of Col-0, *bpm3 bpm5* and *amiR-bpm* INPUTs. Right histogram, ratio of Ub-PP2CA / INPUT PP2CA obtained after quantification of the PD p62 or INPUT samples, respectively. Data are means of two independent experiments ± SD. Asterisk indicates $p < 0.05$ (Student's t test) when comparing data of *bpm3 bpm5* and *amiR-bpm* with Col-0 in the same conditions

5. 50 mL Erlenmeyer flasks (base diameter 64 mm).
6. 10 mM ABA solution: Dissolve 26.4 mg of (+)- cis, trans-abscisic acid in 1 mL of 50 mM Tris–HCl pH 8.5–9.0. Bring the volume to 10 mL with distilled water. Store at −20 °C.
7. 5 mM MG-132: Dissolve 25 mg of MG-132 in 10.5 mL of DMSO. Store at −20 °C.
8. Forceps.
9. Scissors.
10. Mortar.

11. *Arabidopsis thaliana* seeds (Col-0 ecotype).
12. Homozygous *Arabidopsis thaliana* mutant seeds (*bpm3 bpm5* and *amiR-bpm*).

2.2 Affinity Purification of Ubiquitinated Proteins Using p62-Agarose

1. Protein Extraction Lysis Buffer: 150 mM NaCl, 50 mM Tris–HCl pH 8, 1% Triton X-100, Anti-proteases cocktail (Complete mini EDTA free, 1 tablet per 10 mL), 3 mM DTT, 50 μM MG-132, 10 nM ubiquitin aldehyde, 10 mM *N-ethylmaleimide* (NEM) (*see* **Note 1**).
2. 5× Laemmli Buffer: 0.25 M Tris–HCl pH 6.8, 10% sodium dodecyl sulfate (SDS), 0.5 M DTT, 30% (w/v) sucrose, 0.5% (w/v) bromophenol blue, 20% glycerol, 10% β-mercaptoethanol.
3. Bradford, Protein assay Dye Reagent.
4. 1 mg/mL bovine serum albumin (BSA) in water.
5. p62-derived UBA domain agarose immobilize (Enzo) (*see* **Note 2**).
6. Empty unmodified agarose.
7. Spectrophotometer.
8. Tube rocker.
9. Centrifuge.
10. 1.5 mL Eppendorf tubes.

2.3 Analysis by Immunoblotting

1. 10% running gel: 4.1 mL of H_2O, 3.3 mL of 30% Acrylamide/bis (37.5:1), 2.5 mL of 1.5 M Tris–HCl pH 8.8, 100 μL of 10% SDS, 10 μL of TEMED, 100 μL of 10% Ammonium persulfate (APS).
2. Stacking gel: 6.1 mL of H_2O, 1.3 mL of 30% Acrylamide/bis (37.5:1), 2.5 mL of 0.5 M Tris–HCl pH 6.8, 100 μL of 10% SDS, 10 μL of TEMED, 100 μL of 10% APS.
3. 10× Running buffer: 0.25 M Tris, 1.92 M glycine, 1% SDS, dissolve in H_2O.
4. Prestained molecular weight marker SDS-PAGE Standard.
5. Blotting system.
6. Towbin transfer buffer: 25 mM Tris, 192 mM glycine, 20% methanol. Store at 4 °C.
7. PVDF membranes (9 cm × 6 cm).
8. Whatman paper (9 cm × 6 cm).
9. Methanol.
10. 10× TBS: 0.2 M Tris pH 7.5, 1.5 M NaCl. Adjust to pH 7.4.
11. TBS-T: 20 mM Tris pH 7.5, 150 mM NaCl, 0.1% Tween 20.
12. Milk powder.

13. Primary antibodies: α-P4D1 (α-Ub, Santa Cruz Biotechnology, sc-8017), α-E2663 (α-PP2CA, *see* **Note 3**), α-ACTIN.
14. Secondary antibodies: α-rabbit-horseradish-peroxidase, α-mouse-horseradish-peroxidase.
15. ECL™—Western Blotting Detection reagent.
16. Luminescent Image Analyzer (e.g., Fujifilm or equivalent).
17. Stripping buffer: 20 mL of 10% SDS, 12.5 mL of 0.5 M Tris–HCl pH 6.8, 67.5 mL of H_2O, 0.8 mL of β-mercaptoethanol.
18. Shaker.

2.4 Quantification of the Protein Signal

1. NIH Image software ImageJ Fiji (https://imagej.nih.gov/ij/).
2. Spreadsheet program.

3 Methods

3.1 Plant Material and Growth Conditions

1. Surface sterilize Arabidopsis seeds with 70% EtOH +0.1% Triton for 5 min and absolute EtOH for 1 min, dry under laminar hood and plate the sterilized seeds on vertical MS plates. For this experiment we use Col-0 as our control and two mutant lines based on Col-0 (*bpm3 bpm5* and *amiR-bpm*).
2. Stratify the seeds at 4 °C for 2 days and then let them grow for 3 days at 23 °C and light intensity at 120–150 μmol/m^{2}s with 16 h light/8 h dark photoperiod.
3. Autoclave 50 mL Erlenmeyer Flasks with 3 mL of liquid MS medium each (*see* **Note 4**).
4. Transfer 10 plants of each genotype at the same growing stage from the vertical MS plates to different Erlenmeyer Flask with the help of long forceps.
5. Let them grow in the Erlenmeyer Flask under the same conditions, 23 °C and light intensity at 120–150 μmol/m^{2}s with 16 h light/ 8 h dark photoperiod (*see* **Note 5**).
6. After 7 days remove the remaining liquid MS that has not been absorbed by the plants and add 3 mL of fresh liquid MS supplemented with 50 μM ABA and 50 μM MG-132. Once more, let the Erlenmeyer Flask under the same conditions, 23 °C and light intensity at 120–150 μmol/m^{2}s with 16 h light–8 h dark photoperiod (*see* **Note 6**).
7. After 3 h, take the plants out of the Erlenmeyer flask and cut the leaves and the stem and discard them, leaving only the roots (*see* **Note** 7).
8. Powder the roots with liquid nitrogen and the help of a mortar (*see* **Note 8**).

3.2 Affinity Purification of Ubiquitinated Proteins Using p62-Agarose

1. Mix 200 mg of freeze powder with 600 μL of precooled lysis buffer on an Eppendorf tube.
2. Vortex the samples and centrifuge immediately, 11,000 × *g* for 10 min at 4 °C, and collect the supernatant in a new Eppendorf tube (*see* **Note 9**).
3. Separate an aliquot of the supernatant (INPUT), mix 100 μL with 25 μL of 5× Laemmli buffer. Boil the INPUT for 10 min at 95 °C (*see* **Note 10**). Centrifuge at 11,000 × *g* for 1 min and cool on ice.
4. Quantify the concentration of proteins in the extract. For this, dilute the samples 5 times (5 μL of sample + 20 μL of lysis buffer) (*see* **Note 11**). Add 1 μL of the dilute samples to 500 μL of Bradford reagent. Prepare BSA standards (0.5 μg, 1 μg, 1.5 μg, 2 μg, 2.5 μg) to do a concentration curve and also add each of them to 500 μL of Bradford reagent. Measure the absorbance of the BSA standards and the samples at 595 nm with the help of a spectrophotometer.
5. Prepare two Eppendorf tubes for each genotype with 500 μL of either Ub-binding p62 agarose or empty agarose (our negative control). Wash them with 1 mL of lysis buffer and centrifuge at 11,000 × *g* for 1 min at 4 °C. Remove supernatant and cool on ice.
6. Add 300 μg of protein from the test sample to both Eppendorf tubes. Let the tubes in a scientific tube rocker for 2 h at 4 °C.
7. Wash the samples 3 times with 1 mL of lysis buffer. Centrifuge at 11,000 × *g* for 1 min at 4 °C and remove the supernatant after each wash (*see* **Note 12**).
8. Release bound proteins with 200 μL of 2× Laemmli buffer prewarmed at 95 °C.
9. Centrifuge at 11,000 × *g* for 10 min, and collect the supernatant. Boil the samples for 10 min at 95 °C (*see* **Note 10**). Centrifuge at 11,000 × *g* for 1 min and cool on ice.

3.3 Analysis by Immunoblotting

1. Prepare two identical SDS-page gels (10%). Load the three samples of each genotype in each gel: 10 μg of INPUT, 40 μL of eluted p62 agarose and 40 μL of eluted empty agarose. Include one well for a prestained molecular weight marker.
2. Complete the assembly of the gel unit, cover with running buffer and connect to a power supply.
3. Run the gels at 120 V for ~2 h.
4. When the dye fronts are about to run off the gel, stop it.
5. Get ready four Whatman papers and one PVDF membrane for each gel.

6. Incubate the membrane for 1 min in methanol.
7. Disassemble the gel unit and separate the two glasses plates. Cut off the stacking gel and recover the running gel.
8. Wash the gel, the membrane and the Whatman papers in Towbin transfer buffer.
9. Set up the transfer unit in the following order: Cathode core (−), sponge pad, 2 Whatman papers, gel, membrane, 2 Whatman papers, sponge pad, and anode core (+).
10. Transfer the samples overnight (ON) at 12 V at 4 °C (*see* **Note 13**).
11. When the transfer is finished, disassemble the cassette. Use tweezers to move the membranes.
12. From this point on, perform all wash and incubation steps on a shaker.
13. Incubate the membranes for several hours in 20 mL of TBS-T with 5% milk at room temperature (R.T).
14. Incubate the membranes with the primary antibodies ON at 4 °C. One membrane with α-P4D1 (α-Ub) 1:10,000 and the other with α-E2663 (α-PP2CA) 1:2,000 (*see* **Note 14**).
15. Wash the membranes 3 times in TBS-T for 10 min at R.T.
16. Incubate the membranes with the secondary antibody for 1 h at R.T, α-mouse 1:5,000 for the α-Ub and α-rabbit 1:5,000 for the α-PP2CA.
17. Wash the membranes 3 times in TBS-T for 20 min at R.T.
18. Prepare the ECL solution by mixing both components in the ratio 1:1.
19. Incubate the membranes in ECL solution for 5 min and reveal them (*see* **Note 15**).
20. Capture images using the Luminescent Image Analyzer (Fig. 2a). Exposure time depends on the intensity of the signal (*see* **Note 16**).
21. Once you have achieved a satisfactory result, wash the membrane incubated with α-PP2CA with stripping buffer for 5 min at R.T (*see* **Note 17**).
22. Wash the membrane three times with TBS for 5 min, and three more times with TBS-T for 10 min at R.T.
23. Incubate the membrane in 20 mL of TBS-T with 5% milk for 1 h at R.T.
24. Incubate the membrane with α-ACTIN 1:5,000 for 1 h at R.T.
25. Wash the membrane 3 times in TBS-T for 10 min at R.T.
26. Incubate the membrane with the secondary antibody α-rabbit for 1 h at R.T.

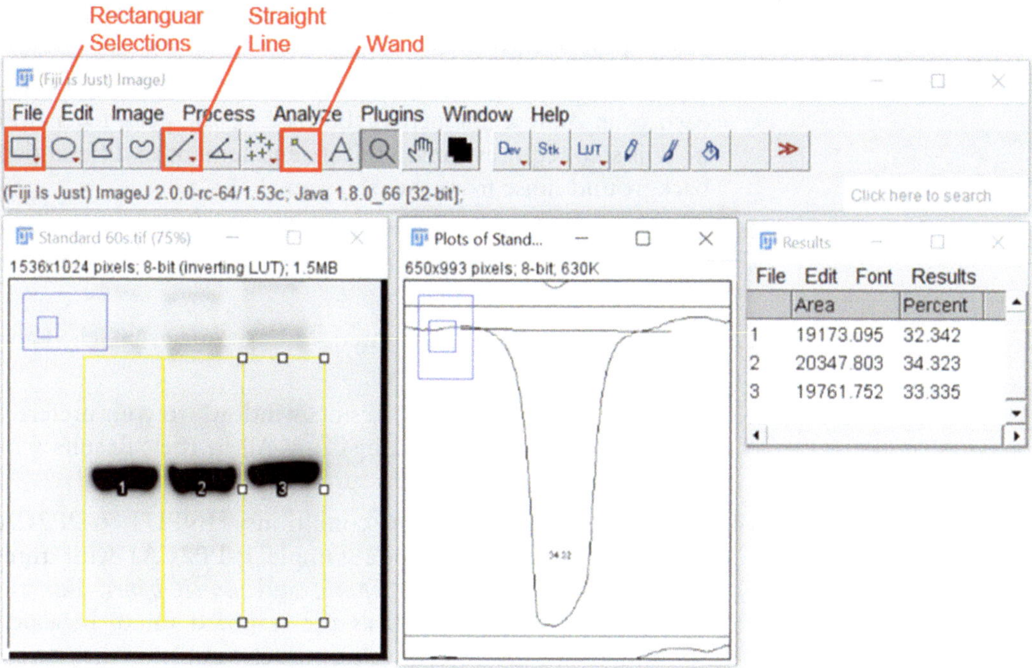

Fig. 3 Representation of the program Image J. Main menu (top) shows tools "Rectangular Selections," "Straight Line," and "Wand" highlighted in red. Open image (left) shows the signal obtained after immunoblotting; yellow rectangles indicate the areas where the plots are performed. Plots panel (central) shows the line to separate the protein signal from background noise. Results window (right) shows the quantification

27. Wash the membrane 3 times in TBS-T for 20 min at R.T.
28. Repeat **steps 18–20** (*see* **Note 18**).

3.4 Quantification of the Protein Signal

1. To analyze the images use ImageJ (Fig. 3).
2. Load the images for the α-PP2CA and α-ACTIN incubated membrane by going to File>Open.
3. If your image is not already grey-scale go to Image>Type>8-bit (*see* **Note 19**).
4. Choose the "Rectangular Selections" tool from the ImageJ toolbar. Draw a vertically long rectangle around the first lane (Fig. 3, yellow rectangle) (*see* **Note 20**).
5. Press the "1" key or go to Analyze>Gels>Select First Lane to set the rectangle in place. The first lane will now be highlighted and have a 1 in the middle of it.
6. Use your mouse to hold the rectangle and move it to the next lane.
7. Press the "2" key or go to Analyze>Gels>Select Next Lane.
8. Continue doing **steps 6** and 7 until all lanes are highlighted (Fig. 3).

9. Press the "3" key or go to Analyze>Gels>Plot Lanes to draw a profile plot of each lane.
10. Choose the "Straight Line" tool from the ImageJ toolbar. For each peak draw a line across the base of the peak to enclose it. This step requires some subjective judgment to differentiate background noise from the peak signal.
11. Choose the "Wand" tool from the ImageJ toolbar. Click with your mouse in each previously enclosed peak from top to bottom.
12. Whit all the peaks highlighted, go to Analyze>Gels>Label Peaks.
13. Move the values from the "Results window" to your preferred spreadsheet by selecting Edit>Copy All in the "Results window." Paste the values into a spreadsheet.
14. There are three different conditions (INPUT α-PP2CA, INPUT α-ACTIN, and p62 sample α-PP2CA) with three peaks each (Col-0, *bpm3 bpm5*, and *amiR-bpm*). For the three conditions use Col-0 as the standard value. In a new column, divide the percent value of each sample by the percent value of the standard (Col-0) to obtain the relative density.
15. Calculate the adjusted density signal of each respective sample by dividing the INPUT α-PP2CA relative density by the INPUT α-ACTIN relative density. Represent in a bar chart (Fig. 2b).
16. Finally divide the p62 sample α-PP2CA relative density by the INPUT α-PP2CA adjusted density signal and represent in another bar chart (Fig. 2b).

4 Notes

1. Our standard lysis buffer does not include ubiquitin aldehyde or NEM. Both reagents inhibit deubiquitination. The NEM solution has to be prepare fresh (12.5 mg of NEM dissolved in 250 μL of absolute EtOH) and neither one can be autoclaved. Store the ubiquitin aldehyde at −20 °C.
2. The resin is supplied as a 50% slurry in 25 mM Tris–HCl, pH 7.5, containing 0.01% sodium azide and can be store at the fridge (4 °C). It is sold in aliquots ready to do one experiment, however, larger quantities can be bought contacting directly with the manufacturer.
3. The α-PP2CA antibody used in this experiment is a self-made antibody reported in [7, 8].
4. Three Erlenmeyer flasks for each genotype are recommended, although material obtained from one flask is usually enough for

the assay. For experiments with a high number of genetic backgrounds 6-multiwell plates can be used instead of flasks. In this case, simply add 1 mL sterile MS medium to each well. The plants should not be submerged in the MS.

5. Check medium remaining in the flasks, if required add 3 mL of MS medium under sterile conditions.
6. From this point on, sterile conditions are no longer required.
7. Take samples out on paper to dry, and then excise the leaves and shoot with the help of forceps and scissors. Avoid using leaves and shoots for the experiment because the RuBisCO protein present in them has a similar molecular weight to PP2CA and it is so abundant that it can mask the signal of the α-PP2CA antibody. Another option is to grow the plants in dark and use the whole plant, which will not have RuBisCO.
8. Check root samples are dry before freezing.
9. Most protocols for lysis of plant tissue recommend incubation in lysis buffer for several minutes; however, in our hands, protein extraction is likewise effective or even more if we proceed directly to centrifugation.
10. Use safe-lock tubes or put something on top of them, otherwise they could open.
11. Protein concentration of the samples is usually too high for direct measurement; therefore, it is important to dilute the samples.
12. Work at 4 °C.
13. ON transfer serves to maximize the amount of high-molecular-mass proteins that are transferred into the membrane. This is not required for proteins with a molecular mass lower than ~50 kDa. In this case, transfer at 120 V for 90 min is possible, however, extra refrigeration should be included into the transfer unit.
14. Only the α-PP2CA antibody require ON incubation.
15. Place the membrane with the side that was in contact with the gel toward the ECL solution.
16. For the membrane incubated with α-Ub, ~1 s exposition should be enough to have a clear signal. For the membrane incubated with α-PP2CA, expose for 1–2 min.
17. The membrane incubated with α-PP2CA is reused to compare the PP2CA signal to the ACTIN signal in the same conditions.
18. ACTIN serves as a loading control, if previous steps were done correctly all bands will have a similar intensity; ~1 s exposition should be enough to have a clear signal.

19. Save your images in the Luminescent Image Analyzer directly in 8-bit format.
20. If the rectangle is horizontally long, the program will generate inverted plots, making them more difficult to read.

Acknowledgments

J.J. was supported by a Formación de Personal Investigador contract (BES-2015-073784) from Ministerio de Ciencia, Innovacion y Universidades (MICIU). Work in P.L.R.'s laboratory was supported by the MICIU and Fondo Europeo de Desarrollo Regional (grant BIO2017-82503-R).

References

1. Cutler C, Rodriguez PL, Finkelstein RR et al (2010) Abscisic acid: emergence of a core signaling network. Annu Rev Plant Biol 61:651–679
2. Moreno-Avero M, Yunta C, Gonzalez-Guzman M et al (2017) Structure of ligand-bound intermediates of crop ABA receptors highlights PP2C as necessary ABA co-receptor. Mol Plant 10(9):1250–1253
3. Park SY, Fung P, Nishimura N et al (2009) Abscisic acid inhibits type 2C protein phosphatases via the PYR/PYL family of START proteins. Science 324(5930):1068–1071
4. Ma Y, Szostkiewicz I, Korte A et al (2009) Regulators of PP2C phosphatase activity function as abscisic acid sensors. Science 324(5930):1064–1068
5. Santiago J, Rodrigues A, Saez A et al (2009) Modulation of drought resistance by the abscisic acid receptor PYL5 through inhibition of clade A PP2Cs. Plant J 60(4):575–588
6. Kong L, Cheng J, Zhu Y et al (2015) Degradation of the ABA co-receptor ABI1 by PUB12/13 U-box E3 ligases. Nat Commun 6:8630
7. Wu Q, Zhang X, Peirats-Llobet M et al (2016) Ubiquitin ligases RGLG1 and RGLG5 regulate abscisic acid signaling by controlling the turnover of phosphatase PP2CA. Plant Cell 28(9):2178–2196
8. Julian J, Coego A, Lozano-Juste J et al (2019) The MATH-BTB BPM3 and BPM5 subunits of Cullin3-RING E3 ubiquitin ligases target clade A PP2Cs for degradation. Proc Natl Acad Sci U S A 116(31):15725–15734
9. Belda-Palazon B, Julian J, Coego A et al (2019) ABA inhibits myristoylation and induces shuttling of the RGLG1 E3 ligase to promote nuclear degradation of PP2CA. Plant J 98(5):813–825
10. Lechner E, Leonhardt N, Eisler H et al (2011) MATH/BTB CRL3 receptors target the homeodomain-leucine zipper ATHB6 to modulate abscisic acid signaling. Dev Cell 21(6):1116–1128
11. Rubio S, Rodriguez A, Saez A et al (2009) Triple loss of function of protein phosphatases type 2C leads to partial constitutive response to endogenous abscisic acid. Plant Physiol 150(3):1345–1355
12. Vlad F, Rubio S, Rodrigues A et al (2009) Protein phosphatases 2C regulate the activation of the Snf1-related kinase OST1 by abscisic acid in Arabidopsis. Plant Cell 21(10):3170–3184
13. Umezawa T, Sugiyama N, Mizoguchi M et al (2009) Type 2C protein phosphatases directly regulate abscisic acid-activated protein kinases in Arabidopsis. Proc Natl Acad Sci U S A 106(41):17588–17593
14. Takahasi Y, Zhang J, Hsu PK et al (2020) MAP3Kinase-dependent SnRK2-kinase activation is required for abscisic acid signal transduction and rapid osmotic stress response. Nat Commun 11(1):12
15. Lin Z, Li Y, Zhang Z et al (2020) A RAF-SnRK2 kinase cascade mediates early osmotic stress signaling in higher plants. Nat Commun 11(1):613
16. Lozano-Juste J, Alrefaei AF, Rodriguez PL (2020) Plant osmotic stress signaling: MAPKKKs meet SnRK2s. Trends Plant Sci 25(12):1179–1182
17. Wang X, Guo C, Peng J et al (2019) ABRE-BINDING FACTORS play a role in the

feedback regulation of ABA signaling by mediating rapid ABA induction of ABA co-receptor genes. New Phytol 221(1):341–355

18. Clausen TH, Lamark T, Isakson P et al (2010) p62/SQSTM1 and ALFY interact to facilitate the formation of p62 bodies/ALIS and their degradation by autophagy. Autophagy 6(3): 330–344
19. Pridgeon JW, Geetha T, Wooten MW (2003) A method to identify p62's UBA domain interacting proteins. Biol Proced Online 5:228–237
20. Bueso E, Ibanez C, Sayas E et al (2014) A forward genetic approach in Arabidopsis thaliana identifies a RING-type ubiquitin ligase as a novel determinant of seed longevity. Plant Sci 215-216:110–116

Chapter 5

Microscopic Imaging of Endosomal Trafficking of ABA Receptors

Borja Belda-Palazón and Pedro L. Rodriguez

Abstract

The abscisic acid (ABA) is a key hormone for stress tolerance. The balance between growth/development and stress responses is crucial for the optimal course of plant life meaning that plants need to control the timing and extent of ABA pathway activation. In this regard, protein turnover regulation by means of both the ubiquitin–proteasome system (UPS) and non-26S proteasome endomembrane trafficking pathways, plays a critical role in the regulation of ABA signaling activation and deactivation. Over the last few years, the ubiquitination of ABA receptors PYRABACTIN RESISTANCE1 (PYR1)/PYR1-LIKE (PYL)/REGULATORY COMPONENTS OF ABA RECEPTORS (RCAR) at the plasma membrane by the RING between RING fingers (RBR)-type E3 ligase RING FINGER OF SEED LONGEVITY1 (RSL1) triggering their internalization through the clathrin-mediated endocytosis (CME) pathway, followed by their endosomal trafficking and delivery to the vacuole for degradation, was reported. For this process, the direct role of some components of the endosomal sorting complex required for transport (ESCRT) machinery, that is, FYVE DOMAIN-CONTAINING PROTEIN 1 (FYVE1)/FYVE DOMAIN PROTEIN REQUIRED FOR ENDOSOMAL SORTING 1 (FREE1) and VACUOLAR PROTEIN SORTING23A (VPS23A) members of ESCRT-I complex, and ALG-2 INTERACTING PROTEIN-X (ALIX) associated protein of ESCRT-III, was reported. In this chapter, we will detail two methods for imaging endosomal trafficking of ABA receptor proteins by confocal microscopy: (a) colocalization of GFP-PYL4 (also known as RCAR10) and CLATHRIN LIGHT CHAIN 2 (CLC2)-mOrange in clathrin-coated vesicles in *Nicotiana benthamiana* leaf cells and (b) localization of GFP-PYL4 into Wortmannin (WM)-enlarged late endosomes in *Arabidopsis thaliana* root cells.

Key words Abscisic acid receptors, PYR/PYL/RCAR, CLC2, RSL1, Clathrin-mediated endocytosis, Endosomal trafficking, Vacuolar degradation, ESCRT, Wortmannin

1 Introduction

In response to adverse environmental cues such as drought, heat, and cold stress or high salinity, ABA levels in plants rise to adapt their growth and development accordingly [1]. ABA signaling promotes plant responses through a complex signal transduction pathway that starts with the ABA perception by the PYR/PYL/RCAR ABA receptors, which constitute a multigene family of

Takuya Yoshida (ed.), *Abscisic Acid: Methods and Protocols*,
Methods in Molecular Biology, vol. 2462, https://doi.org/10.1007/978-1-0716-2156-1_5,

14 members [2]. ABA receptors are soluble proteins localized both at the cytosol and the nucleus, but they can be also found as peripheral membrane proteins through interaction with the C2-domain ABA-related (CAR) proteins [3, 4].

Excessive duration of hormone activation can negatively impact plant development. Hence, once stress dissipates, plants put into operation diverse mechanisms of negative feedback for desensitization of ABA signaling [5]. Regulated protein turnover plays a critical role for hormone signaling [6], and regulation of receptor stability has emerged as a major mechanism for termination of the signal cascade and ABA desensitization [5]. Concerning this, both UPS and non-26S proteasome endomembrane trafficking pathways are involved in the regulation of ABA signaling deactivation [7–13]. Ubiquitination of membrane proteins via K63-linked chains is a signal that triggers endocytosis and cargo sorting through the endosome route [14]. Ubiquitination of PYR1 (also known as RCAR11) and PYL4 ABA receptors at the plasma membrane by the RBR-type E3 ligase RSL1 triggers their internalization through the CME pathway and trafficking to the vacuole for degradation by means of the ESCRT machinery [8, 11]. In this process, ubiquitinated receptors are recognized by FYVE1/FREE1 and VPS23A components of ESCRT-I complex that target ABA receptors for vacuolar degradation [9, 10]. Furthermore, ALIX, associated protein of ESCRT-III that interacts with FYVE1/FREE1 and VPS23A, also binds directly PYL4, PYL5 (also known as RCAR8), PYL8 (also known as RCAR3), and PYL9 (also known as RCAR1) ABA receptors, and mediates sorting to multivesicular bodies (MVBs) for their degradation [12]. These reports suggest that controlled turnover of ABA-receptors at protein level is crucial for the proper control of the ABA signaling pathway [5].

We reported previously that ABA receptors follow endocytic turnover from the plasma membrane to the vacuole. Proteomic analyses of immunocomplexes obtained upon immunoprecipitation of HA-tagged PYR1, PYL4, or PYL8 revealed their association with components of the clathrin-dependent coating and scission machinery, such as clathrin heavy chain (CHC)/clathrin light chain (CLC) subunits and different dynamin-related proteins, respectively [9]. Furthermore, we found some other components are also required for endocytosis, for instance some members of the AP-2 and the TPLATE adaptor complexes, which are recruited at the plasma membrane preceding recruitment of clathrin and dynamin-related proteins [15]. Further evidence of the CME of ABA receptors was achieved by expressing GFP-PYL4 and CLC2-mOrange fusion proteins in *N. benthamiana* leaves and detecting their colocalization in endocytic vesicles near the plasma membrane. Additionally, to investigate the role of FYVE1/FREE1 in the sorting of ABA receptors we expressed GFP-PYL4 in wild-type and *fyve1-3* plants. Treatment with WM, an inhibitor of phosphatidylinositol 3-kinase (PI3-K) that generates enlargement of MVBs and prevacuolar compartments (PVCs), revealed that GFP-PYL4 did not reach enlarged MVBs/PVCs in the *fyve1-3* mutants [9]. In this chapter, we will provide two detailed protocols for the microscopic imaging of endosomal trafficking of ABA receptors. Firstly, the colocalization of GFP-PYL4 with CLC2-mOrange in clathrin-mediated endocytic vesicles in *N. benthamiana* leaf cells, and secondly, the trafficking of GFP-PYL4 through MVBs/PVCs in *A. thaliana* root cells of wild-type or *fyve1-3* plants.

2 Materials

Unless otherwise indicated, prepare all solutions using sterile ultrapure water (18 megOhm, e.g., Milli-Q) and store all the reagents at room temperature.

2.1 Analysis of the Localization of PYL4 in Clathrin-Coated Vesicles

1. *Agrobacterium tumefaciens* strains encoding the appropriate constructions (Table 1).
2. Luria-Bertani medium (LB): 10 g/L tryptone, 5 g/L yeast extract, 10 g/L NaCl. Dissolve 10 g of tryptone, 5 g of yeast extract, and 10 g of NaCl in 700 mL of water. Bring the volume to 1 L with water, autoclave it for 20 min at 120 °C and 1 atm of pressure and stored at room temperature. To supplement 5 mL of LB with 50 μg/mL kanamycin and 25 μg/mL rifampicin add 5 μL of 50 mg/mL kanamycin and 5 μL of 25 mg/mL rifampicin. To supplement 5 mL of LB with 50 μg/mL spectinomycin and 25 μg/mL rifampicin add 5 μL of 50 mg/mL spectinomycin and 5 μL of 25 mg/mL rifampicin.
3. 50 mL polypropylene tubes and tube adapters.
4. Orbital shaker.
5. 1.5 mL methacrylate cuvettes.
6. Cell Density Meter.
7. Refrigerated centrifuge for 50 mL tubes.
8. MES (2-(N-morpholino) ethanesulfonic acid) Buffer: 10 mM $MgCl_2$, 10 mM MES pH 5.6, 200 μM acetosyringone. Dissolve 0.095 g of $MgCl_2$ (MW 95.21) and 0.195 g of MES (MW 195.24) in 80 mL of water. Adjust the pH to 5.6 with KOH, add 100 μL of 200 mM acetosyringone and bring the volume to 100 mL with water. Prepare it fresh just before resuspending the *A. tumefaciens* strains.
9. 1 mL needleless syringes.
10. Controlled-environment growth chamber (*see* **Note 1**).
11. 4-week-old *N. benthamiana* plants (*see* **Note 1**).

Table 1
***A. tumefaciens* strains used in this work and the corresponding proteins expressed from each binary vector**

Protein expression/vector	Reference	Bacterial selection	*A. tumefaciens* strain
GFP-PYL4 (pMDC43)	Belda-Palazon et al. (2016) [9]	Kanamycin (50 μg/mL)	C58C1/pGV2260
		Rifampicin (25 μg/mL)	
proCLC2:CLC2-mOrange	Konopka et al. (2008) [16]	Spectinomycin (50 μg/mL)	C58C1/pGV2260
(*pPZP221B-mO-NOS*)		Rifampicin (25 μg/mL)	
p19 (pCB301)	Belda-Palazon et al. (2012) [17]	Kanamycin (50 μg/mL)	C58C1/pCH32
		Rifampicin (25 μg/mL)	

12. Hole puncher.
13. Microscope slides and glass coverslips.
14. Confocal laser scanning microscope (*see* **Note 2**).
15. ImageJ software (http://rsb.info.gov/ij/) and its PSC Colocalization plug-in (https://nottingham.ac.uk/research/groups/cvl/software/psccolocalisation.aspx).

2.2 Localization of PYL4 into WM-Enlarged MVBs/PVCs

1. Homozygous *A. thaliana 35S::GFP-PYL4* wild-type and *35S::GFP-PYL4 fyve1-3* seeds.
2. 120 × 120 mm squared Petri dishes.
3. 12-well cell culture plates: Growth area 3.5 cm^2 per well.
4. Controlled-environment growth chamber (*see* **Note 1**).
5. pH meter.
6. Autoclave.
7. Stainless-steel precision forceps with fine tips.
8. 1× Murashige and Skoog (1 × MS) medium plates supplemented with 1% sucrose: Dissolve 4 g of Murashige and Skoog including vitamins and MES buffer and 10 g of sucrose in 700 mL of water. Adjust pH to 5.7 with a 1 M KOH solution (*see* **Note 3**) by using the pH meter and bring the volume to 1 L with water. Add 10 g of Phytoagar and sterilize the medium at 120 °C 1 atm for 20 min by autoclaving. Inside the laminar flow hood, pour 50 mL of medium into 120 × 120 mm squared Petri dishes and let the medium solidify. Store the plates at 4 °C before use.

9. 12 wells culture plates containing 0.5 mL of sterile 1×MS + 1% sucrose: Dissolve 4 g of Murashige and Skoog including vitamins and MES buffer, and 10 g of sucrose in 700 mL of water. Adjust pH to 5.7 with a 1 M KOH solution by using the pH meter and bring the volume to 1 L with water. Sterilize it at 120 °C and 1 atm of pressure for 20 min by the autoclave. Add 0.5 mL of liquid 1×MS + 1% sucrose medium into each well just before the drug treatment.
10. 3.3 mM WM stock solution: To obtain firstly the 33 mM WM (1000×), add 354 μL of DMSO to 5 mg of WM (MW = 428.43) and mix well. Afterward, prepare 500 μL of 3.3 mM WM (100×) by mixing 10 μL of 33 mM WM (1000×) with 490 μL of DMSO. Store the stock solutions at −20 °C.
11. 1 mM LysoTracker® Red DND-99: Store at −20 °C and protect from light.
12. 1.5 mL tubes with 1 mL of sterile ultrapure water.
13. Pasteur pipettes and rubber bulbs.
14. Microscope slides and Glass Coverslips.
15. Confocal laser scanning microscope (*see* **Note 2**).
16. ImageJ software (http://rsb.info.gov/ij/).

3 Methods

3.1 Analysis of the Localization of PYL4 in Clathrin-Coated Vesicles

In this protocol, we analyze the colocalization of GFP-PYL4 with CLC2-mOrange in clathrin-coated vesicles at the plasma membrane in *N. benthamiana* leaves. Clathrin is a molecular scaffold for vesicular uptake of cargo at the plasma membrane and defines the CME for plasma membrane–based proteome turnover in eukaryotes [15].

3.1.1 Agroinfiltration of N. benthamiana Leaves

1. Obtain saturated *A. tumefaciens* cultures (3 mL) containing the appropriate construct depicted in Table 1.
2. Grow 5 mL cultures of the *A. tumefaciens* strains encoding the constructs of interest. To do that, in a 50 mL tube inoculate 25 μL of each saturated culture into 5 mL of LB containing the appropriate antibiotics. Incubate the cultures at 28 °C in an orbital shaker (180 rpm) for 14 h.
3. Measure the optical density at 600 nm (OD_{600nm}) of each culture. For this, firstly dilute 1:10 each culture by mixing 100 μL of each culture with 900 μL of LB. Pipette the dilutions into 1.5 mL methacrylate cuvettes and measure the OD_{600nm} in the Cell Density Meter. Then, multiply by 10 the value of OD_{600nm} to obtain the real cellular density of each bacterial culture (*see* **Note 4**).

4. Centrifuge the cultures at 3000 × *g* for 15 min, discard the supernatant, and resuspend the cells in MES buffer to an OD_{600nm} of 1.0 (*see* **Note 5**). Afterward, incubate gently the cultures at room temperature in the orbital shaker at 50 rpm for 3–4 h.
5. Mix one volume (1:1:1) of each *A. tumefaciens* strain: 1 mL of GFP-PYL4, 1 mL of CLC2-mOrange and 1 mL of p19 (*see* **Note 6**) and infiltrate the mixture into the abaxial air space of *N. benthamiana* leaves with a needleless syringe (*see* **Note 7**).
6. Incubate the infiltrated plants in a controlled-environment growth chamber under long day (LD) photoperiod conditions (*see* **Note 1**) for 72 h.

3.1.2 Colocalization Analysis

1. With a hole puncher, obtain sections of the infiltrated leaves and mount them with water onto microscope slides leaving the abaxial side up and the adaxial side down. Cover the preparations with coverslips.
2. Through the confocal microscope, observe the GFP and mOrange signals generated by the expression of GFP-PYL4 and CLC2-mOrange, respectively (*see* **Note 2**). Table 2 describes the conditions used for the imaging.
3. Focus on the plasma membrane of the mesophyll cells and move through the depth (Z-plane) of the cells. Search for circular and/or dotted (vesicle-like) structures decorated by both, GFP and mOrange signals.
4. Once pictures are taken, analyze the colocalization of GFP-PYL4 and CLC2-mOrange in these structures by the ImageJ software and its PSC Colocalization plug-in. To do this, firstly select the Region of Interest (ROI) with the ROI Manager tool in the ImageJ (*see* **Note 8**).
5. Obtain the 2D histogram/scatterplot to observe the correlation between pixel intensities in the image and the linear regression of the data in the scatterplot.
6. Determine automatically the Costes threshold by which the intensity levels above which colocalization for both channels are defined (*see* **Note 9**).
7. Obtain the Pearson's (above the Costes threshold) and Spearman's correlation values (R_P and R_S respectively). Values in the range + 0.4 to 1 indicate colocalization, whereas lower values or negative values indicate lack of colocalization.
8. In this example, Pearson's (R_P = 0.92) and Spearman's (R_S = 0.97) correlation coefficients indicate colocalization of GFP-PYL4 with CLC2-mOrange in clathrin-coated vesicles at the plasma membrane (Fig. 1).

Table 2
CLSM conditions for the visualization of GFP and mOrange fluorescent proteins, and the LysoTracker® Red DND-99 fluorescent probe

Fluorescent protein/probe	Excitation laser (nm)	MBS = dichroic	Spectral detection (nm)
GFP	Argon 488	488	500–530
mOrange	Argon 561	488/561	550–580
LysoTracker® red DND-99	Argon 561	488/561	585–615

MBS main beam splitter

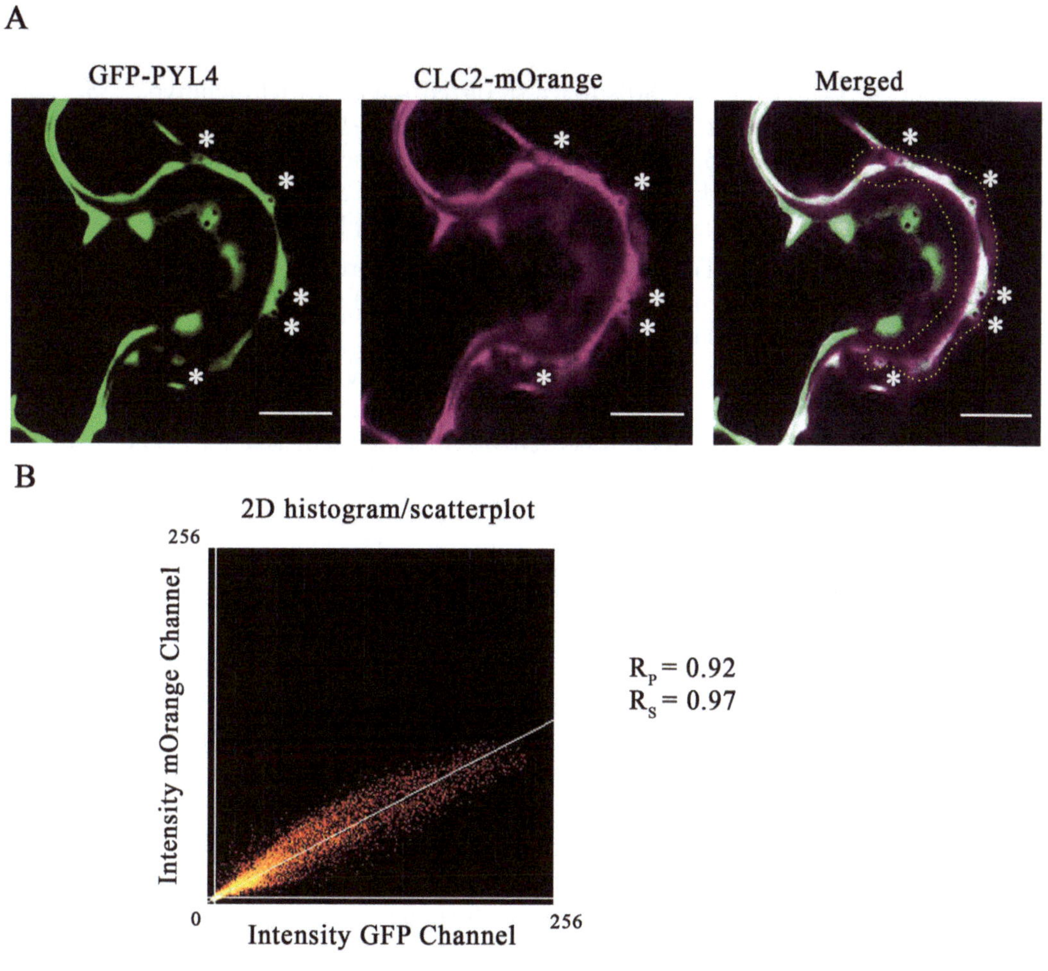

Fig. 1 PYL4 colocalizes with CLC2 in clathrin-coated vesicles at the plasma membrane. (**a**) Confocal images obtained from transiently transformed *N. benthamiana* leaf cells coexpressing both GFP-PYL4 and CLC2-mOrange. Asterisks indicate punctate/globular structures in plasma membrane. Dashed yellow line marks the ROI. (**b**) 2D histogram/scatterplot of the ROI representing the correlation of pixel intensities (0-256) of both channels (GFP and mOrange). In the graph are also represented the linear regression of the data (diagonal white line) and the Costes threshold for both channels (vertical and horizontal white lines). Bars = 30 μm

3.2 Localization of PYL4 into WM-Enlarged MVBs/PVCs

In this protocol we analyze the localization of GFP-PYL4 in MVBs/PVCs which are enlarged by WM. These late endosomes are stained by the acidotropic dye LysoTracker® Red DND-99. Specifically, we study the endosomal trafficking of the ABA receptor PYL4, which is analyzed in *A. thaliana* plants overexpressing GFP-PYL4, either in wild-type or *fyve1-3* genetic background [9]. *fyve1-3* mutants are impaired in the targeting of PYL4 to late endocytic vesicles (MVBs/PVCs) prior to vacuolar degradation [9]. WM is a PI3-K inhibitor that inhibits protein cargo trafficking to the vacuole. Therefore, WM-treated plants are used to obtain enlargement of MVBs/PVCs [18]. Furthermore, the presence of PYL4 in these compartments are investigated using LysoTracker® Red DND-99, an acidophilic probe that labels acidic/lytic compartments [19].

1. Sow 50 surface-sterilized seeds of each genotype (*35S::GFP-PYL4* wild-type and *35S::GFP-PYL4 fyve1-3*) on solid 1×MS plates supplemented with 1% sucrose. Conduct stratification in the dark at 4 °C for 3 days and then grow the seedlings vertically in vitro for 4 days (*see* **Note 1**).
2. Transfer carefully with sterile forceps 5 seedlings to each well of the culture plate containing 0.5 mL of sterile 1×MS + 1% sucrose (*see* **Note 10**).
3. Perform the WM treatments by adding 5 μL of 3.3 mM WM directly into the medium of each well and incubate for 3 h in the controlled-environment growth chamber. The final concentration of WM is nearly 33 μM (*see* **Note 11**).
4. Add 1 μL of 1 mM LysoTracker® Red DND-99 directly into the medium of each well and incubate one additional hour in the growth chamber. The final concentration of LysoTracker® Red DND-99 is nearly 2 μM.
5. Place a drop of sterile water from the 1.5 mL tube onto the center of the microscope slide by the Pasteur pipette.
6. Carefully lay one seedling into the water with the forceps and cover the preparation with the coverslip (*see* **Note 10**).
7. Observe the green (GFP-PYL4) and red (LysoTracker® Red DND-99) fluorescence signals through confocal laser scanning microscopy (CLSM) (*see* **Note 2**). Table 2 describes the conditions used for the imaging.
8. Focus on the region of the root apex covering both, the cells of the distal part of the elongation zone and the cells of the beginning of the grow terminating zone, which is located 700–1000 μm from the root cap junction (RCJ) (*see* **Note 12**).
9. Search for cells presenting rings decorated in red fluorescence corresponding to acidic/lytic endosomes, and analyze carefully the localization of the GFP-PYL4 through the green fluorescence.

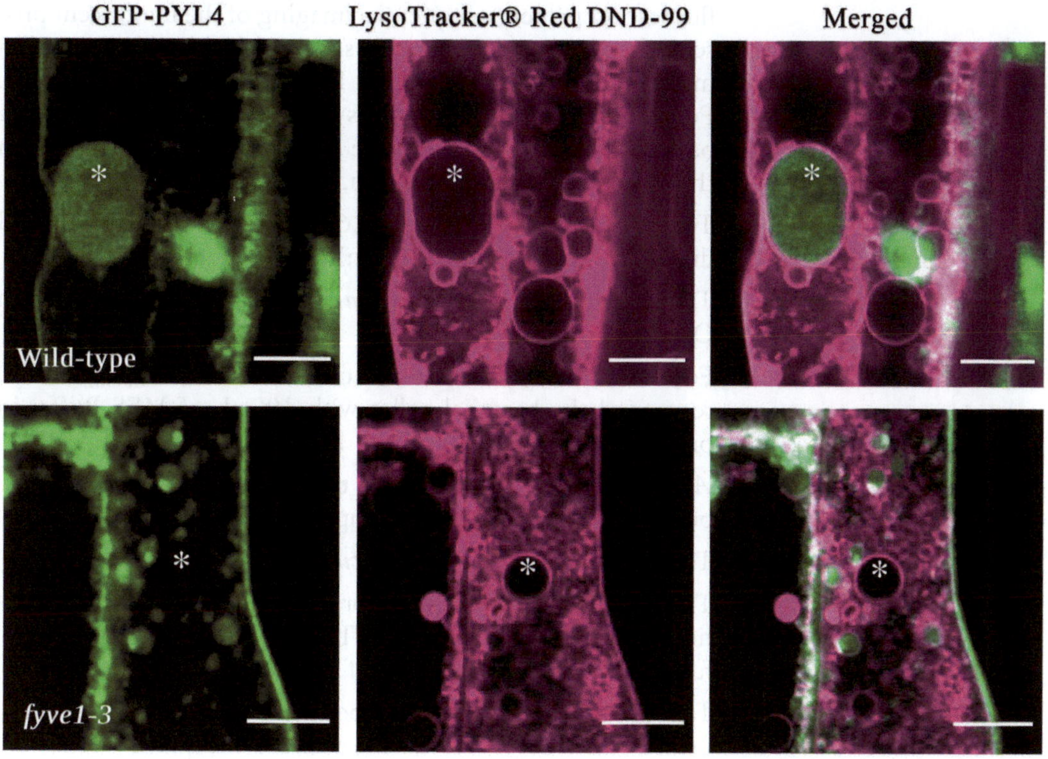

Fig. 2 GFP-PYL4 localizes in WM-enlarged late endosomes (indicated by asterisks) in wild-type seedlings but it is absent in *fyve*-3 mutants. Confocal images obtained from both *A. thaliana* wild-type (upper panels) and *fyve1-3* (lower panels) root cells, overexpressing GFP-PYL4. Seedling were treated with WM and acidic/lytic compartments were decorated by LysoTracker Red DND-99 stainning. Bars = 10 μm

10. In this example, whereas in wild-type seedlings GFP-PYL4 is observed in enlarged late endocytic vesicles (enlarged MVBs/PVCs) stained by LysoTracker Red DND-99, GFP-PYL4 is only present in early acidic/lytic endosomes and absent from enlarged late endocytic compartments in *fyve1–3* (Fig. 2).

4 Notes

1. In this example, both *N. bethamiana* growth and plant in vitro culture are conducted in a controlled-environment growth chamber at 22 °C under LD photoperiod conditions (16 h light–8 h dark photoperiod) at 80–100 μmol m^{-2} s^{-1} of light intensity.
2. In this example, CLSM is performed using a Zeiss LSM 780 AxioObserver.Z1 laser scanning microscope with a C-Apochromat 403/1.20 W corrective water immersion objective lens. As this is involving multicolor detection of two

fluorescent proteins, sequential imaging of the fluorescent proteins is performed using the sequential channel acquisition mode. Pinholes are adjusted to 1 air unit for each wavelength. For the colocalization analyses it is critical that all imaging parameters which influence the intensity of the image remain the same for all data acquisition.

3. To prepare 1 M KOH, weigh 5.61 g of KOH (MW 56.11) and dissolve it in water up to 100 mL.
4. The growth of the *A. tumefaciens* strains should reach an OD_{600nm} around 2.0.
5. For an *A. tumefaciens* culture of 5 mL and $OD_{600nm} = 2.0$, resuspend the bacterial pellet with 10 mL of MES Buffer in order to bring the OD_{600nm} to 1.0.
6. *A. tumefaciens* cells expressing the p19 protein of the tomato bushy stunt virus are used to suppress gene silencing.
7. The final OD_{600nm} of each *A. tumefaciens* strain is 0.33.
8. The ROI Manager is a tool for working with multiple selections in the ImageJ software. This tool allows to outline carefully and manually the region of interest on the original image.
9. The pixels with intensities above the two thresholds have greater than zero correlation whereas the pixels below the thresholds have none or anti correlated intensities.
10. Handle the seedlings meticulously. Do not press excessively.
11. WM treatments have to be previously set empirically for the specific stages and tissues mentioned in this case example.
12. *A. thaliana* root apices have four distinct zones of growth activities: the meristematic zone (up to 150–200 μm from the RCJ), the transition zone (200 μm to 520 μm from the RCJ), the elongation zone (200 to about 520 μm away from the RCJ), and the grow terminating zone [20].

Acknowledgments

We acknowledge S.Y. Bednarek for the *proCLC2:CLC2-mOrange* construct. Work in Dr. Rodriguez' s laboratory was supported by the Ministerio de Ciencia, Innovacion y Universidades, Fondo Europeo de Desarrollo Regional and Consejo Superior de Investigaciones Cientificas (grant BIO2017-82503-R to P.L.R.). B.B-P. was funded by the European Union Horizon 2020 research and innovation programme (grant agreement no. 867426—ABA-GrowthBalance—H2020-WF-2018-2020/H2020-WF-01-2018).

References

1. Chen K, Li GJ, Bressan RA et al (2020) Abscisic acid dynamics, signaling, and functions in plants. J Integr Plant Biol 62(1):25–54
2. Rodriguez PL, Lozano-Juste J, Albert A (2019) Chapter two - PYR/PYL/RCAR ABA receptors. In: Seo M, Marion-Poll A (eds) Advances in botanical research, vol 92. Academic Press, London, pp 51–82
3. Rodriguez L, Gonzalez-Guzman M, Diaz M et al (2014) C2-domain abscisic acid-related proteins mediate the interaction of PYR/-PYL/RCAR abscisic acid receptors with the plasma membrane and regulate abscisic acid sensitivity in Arabidopsis. Plant Cell 26(12): 4802–4820
4. Diaz M, Sanchez-Barrena MJ, Gonzalez-Rubio JM et al (2016) Calcium-dependent oligomerization of CAR proteins at cell membrane modulates ABA signaling. Proc Natl Acad Sci U S A 113(3):E396–E405
5. Ali A, Pardo JM, Yun D-J (2020) Desensitization of ABA-signaling: the swing from activation to degradation. Front Plant Sci 11:379
6. Santner A, Estelle M (2009) Recent advances and emerging trends in plant hormone signalling. Nature 459(7250):1071–1078
7. Irigoyen ML, Iniesto E, Rodriguez L et al (2014) Targeted degradation of abscisic acid receptors is mediated by the ubiquitin ligase substrate adaptor DDA1 in Arabidopsis. Plant Cell 26(2):712–728
8. Bueso E, Rodriguez L, Lorenzo-Orts L et al (2014) The single-subunit RING-type E3 ubiquitin ligase RSL1 targets PYL4 and PYR1 ABA receptors in plasma membrane to modulate abscisic acid signaling. Plant J 80(6): 1057–1071
9. Belda-Palazon B, Rodriguez L, Fernandez MA et al (2016) FYVE1/FREE1 interacts with the PYL4 ABA receptor and mediates its delivery to the vacuolar degradation pathway. Plant Cell 28(9):2291–2311
10. Yu F, Lou L, Tian M et al (2016) ESCRT-I component VPS23A affects ABA signaling by recognizing ABA receptors for endosomal degradation. Mol Plant 9(12):1570–1582
11. Yu FF, Xie Q (2017) Ubiquitination modification precisely modulates the ABA signaling pathway in plants. Trends Plant Sci 22(11): 976–985
12. García-León M, Cuyas L, El-Moneim DA et al (2019) Arabidopsis ALIX regulates stomatal aperture and turnover of abscisic acid receptors. Plant Cell 31(10):2411–2429
13. Fernandez MA, Belda-Palazon B, Julian J et al (2020) RBR-type E3 ligases and the ubiquitin-conjugating enzyme UBC26 regulate abscisic acid receptor levels and signaling. Plant Physiol 182(4):1723–1742
14. Paez Valencia J, Goodman K, Otegui MS (2016) Endocytosis and endosomal trafficking in plants. Annu Rev Plant Biol 67:309–335
15. Gadeyne A, Sánchez-Rodríguez C, Vanneste S et al (2014) The TPLATE adaptor complex drives clathrin-mediated endocytosis in plants. Cell 156(4):691–704
16. Konopka CA, Backues SK, Bednarek SY (2008) Dynamics of Arabidopsis dynamin-related protein 1C and a clathrin light chain at the plasma membrane. Plant Cell 20(5):1363–1380
17. Belda-Palazón B, Ruiz L, Martí E et al (2012) Aminopropyltransferases involved in polyamine biosynthesis localize preferentially in the nucleus of plant cells. PLoS One 7(10): e46907
18. Robinson DG, Jiang L, Schumacher K (2008) The endosomal system of plants: charting new and familiar territories. Plant Physiol 147(4): 1482–1492
19. Nodzynski T, Feraru MI, Hirsch S et al (2013) Retromer subunits VPS35A and VPS29 mediate prevacuolar compartment (PVC) function in Arabidopsis. Mol Plant 6(6):1849–1862
20. Verbelen J-P, Cnodder TD, Le J et al (2006) Root apex of *Arabidopsis thaliana* consists of four distinct zones of growth activities: meristematic zone, transition zone, fast elongation zone, and growth terminating zone. Plant Signal Behav 1(6):296–304

Chapter 6

ABA Transport Assay in Plant Single-Cell System

Joohyun Kang

Abstract

Compare with long history of the abscisic acid (ABA) study, our knowledge of ABA transporters is quite recent. This is due to at least one reason: ABA is a weak acid, and thus, it exists in either protonated form or in anionic form depending on the surrounding pH relative to its pK_a value. Because the protonated form of ABA can permeate the cell membrane, it would foreclose a specific uptake transporter of ABA. Notwithstanding this theoretical base, ABA transporters belonging to different protein families have been reported a decade ago, steadily. A critical point of the identification of novel ABA transporters is to prove their transport activity. To do this, heterologous expression system is considered first as a facility of transport activity analysis. However, it is difficult to overexpress membrane proteins in their functional state in heterologous system. They have the tendency to aggregate and produce inclusion bodies caused by mistargeting. Thus, in this chapter, I describe the method of ABA transport assay based on single-cell system originated from plant tissues.

Key words Abscisic acid, Protoplasts, BY-2 cells, ABA transporter, Transport activity analysis, Uptake assay, Isotope-labeled ABA

1 Introduction

Abscisic acid (ABA) biosynthesis occurs in limited types of cells, not in all kinds of cells [1]. The primary synthesis place is vascular tissues due to high contribution of xylem parenchyma cells and phloem companion cells in Arabidopsis [2, 3]. It supports ABA redistribution within a leaf not translocated from root to shoot under water stress condition [4]. Guard cells, the main acting site of ABA as a stoma opening and closing movement regulator, are not connected to mesophyll or phloem parenchyma cells via plasmodesmata and ABA signal perception mainly occurs in cytosol by cytoplasm-localized ABA receptor, PYRABACTIN RESISTANCE1 (PYR1)/PYR1-LIKE (PYL)/REGULATORY COMPONENTS OF ABA RECEPTOR (RCAR) protein family [5–7]. It means ABA is required to cross membranes for at least two functions:

Takuya Yoshida (ed.), *Abscisic Acid: Methods and Protocols*,
Methods in Molecular Biology, vol. 2462, https://doi.org/10.1007/978-1-0716-2156-1_6,

exporting from vascular parenchyma cells and uptake into guard cells, for stomatal regulation.

Since ABA is a weak acid (pKa = 4.7), it exists in either protonated form (uncharged form, e.g., ABAH) or in anionic forms (charged form, e.g., ABA^{-}) depending on the surrounding pH. Anionic form of ABA command a majority in cytosol relatively alkaline compartment while ABA exists as a protonated form in apoplastic stream, a relatively acidic compartment of plants. Because the protonated form of ABA easily permeate the cell membrane, it was historically assumed that this passive diffusion process would obviate a requirement for specific ABA uptake protein [8]. However, an increase in pH in xylem sap and apoplast has been observed in many plant species under water stress conditions [9], which enforces a requirement of ABA uptake transporter also, because the population of anionic form of ABA in apoplast would increase with increasing pH. Attesting to this, several ABA transporters have been reported [10].

The ABA transporters that have been identified to date belong to several different protein families—the ABC (ATP-binding cassette) superfamily, the NPF (nitrate transporter1/peptide transporter family) family, and the DTX/MATE (detoxification efflux carriers/multidrug and toxic compound extrusion) family in Arabidopsis. The large number of loss-of-function ABA transporter mutants showed the relatively marginal phenotypes. It indicates that there is considerable redundancy among these proteins. In this context, a critical point of the identification of novel ABA transporters is to prove their transport activity.

Two different plasma membrane–localized ABC transporters were identified as ABA transporters, first. Transporters belonging to ABC protein family occupy the majority of ABA transporters ever since. AtABCG25 is localized predominantly to the vasculature of leaves, the main ABA biosynthetic site, and ABA export activity has been examined using membrane vesicles isolated from insect cells overexpressing AtABCG25 [11]. AtABCG40 has been characterized as an ABA importer in guard cells. The ABA import activity of AtABCG40 has been studied using yeast and tobacco suspension culture cells, BY-2 cells expressing AtABCG40 as well as mesophyll protoplast isolated from loss-of-function ABCG40 transporter mutants [12].

One of the low-affinity nitrate transporters, NRT1.2/NFP4.6, exhibited ABA transport activity when expressed in yeast, insect vesicles [13], and oocyte [14]. Using *E. coli* and *Xenopus* as heterologous expression systems, as well as wild-type and *dtx50* Arabidopsis protoplasts, it has been revealed AtDTX50 convincingly shows ABA exporting activity. Guard cell–localized AtDTX50 contributed to fine tuning the regulation of stomatal aperture by releasing ABA from guard cells [15].

All the work I have described so far used Arabidopsis as a model plant. Much less is identified in other plant species. Lr34, ABC protein in wheat, revealed its ABA transport activity in Lr34 expressing yeast cells [16]. By measuring fluorescence resonance energy transfer in yeast expressing OsPM1, a membrane protein belonging to the AWPM-19 family in rice, it has been shown to exhibit ABA uptake activity [17]. MtABCG20, ABC protein from *Medicago truncatula* that was identified as an ABA exporter. By introducing MtABCG20 into BY-2 cells, authors observed that ABA was released much faster by these cells than by those transformed with an empty vector. Furthermore, ABA transport activity also was observed in vesicles isolated from these cells [18]. Recent study reported a homolog of AtABCG25, ABA efflux transporter, in *Brachypodium distachyon*, named BdABCG25, excreted ABA more efficiently from Sf9 insect cells expressing *BdABCG25* [19].

To prove their transporting activity of ABA, ABA transporters reported to date often utilized heterologous expression system (e.g., yeast, insect cells, oocytes) as a facility of transport activity analysis. However, it is difficult to overexpress membrane proteins in their functional state in heterologous system. They have the tendency to aggregate and produce inclusion bodies caused by misfolding and/or mistargeting of introduced protein. It also failed to measure transport activity with the absence of interaction partner or suitable post-translational modification provided from their own system.

Thus, in this chapter, I describe how to perform ABA transport assay based on single-cell system originated from plant tissues. I will describe how to isolate Arabidopsis mesophyll protoplast, generate ABA transporter expressing suspended BY-2 cell lines, and report the transport procedure using them.

2 Materials

2.1 ABA Transport Assay with Arabidopsis Protoplasts

Prepare all solutions using ultrapure water (prepared by purifying deionized water, to attain a sensitivity of 18 MΩ-cm at 25 °C) and analytical grade reagents.

2.1.1 Mesophyll Protoplast Isolation

1. Seven-week-old Arabidopsis plants grown in short-day (8 h light at 22 °C/16 h dark at 20 °C) on autoclaved soil in environment-controlled chamber.
2. Plant growth chamber.
3. P500 sandpaper.
4. Fine precise forceps, straight scissors for Arabidopsis.

5. MCP buffer: 500 mM sorbitol, 1 mM $CaCl_2$, 10 mM 2-(N-morpholino)ethanesulfonic acid (MES). Adjust pH to 5.8 with 0.1 M KOH. Keep at room temperature after autoclaving.
6. Plastic or glass petri dish: 145 mm diameter × 17 mm height or similar size.
7. Equipment for suction.
8. Digestion buffer: 0.4 M Mannitol, 20 mM KCl, 20 mM MES, 10 mM $CaCl_2$. Adjust pH to 5.8 with 0.1 M KOH. Keep at room temperature after autoclaving. Freshly add 1% cellulose R10, 0.4% macerozyme R10, 0.1% bovine serum albumin (BSA) (*see* **Note 1**).
9. Analogue reciprocal shaker or Digital low speed reciprocal shaker.
10. Sieve with 100 μm nylon mesh.
11. 1 M $CaCl_2$: Keep at room temperature after autoclaving.
12. Trimmed 1 mL pipette tips: 4–5 mm cut off from narrow end.
13. 50 mL round-bottom centrifuge glass tubes.
14. Benchtop swing-bucket centrifuge.
15. 21% sucrose solution.
16. Falcon test tube with snap cap, 14 mL round bottom.
17. Hemocytometer and optical microscope.
18. Pipette aid with 25 mL tips.

2.1.2 ABA Accumulation/Uptake Assay

1. Percoll solution: 500 mM sorbitol, 1 mM $CaCl_2$, 10 mM MES in Percoll centrifugation media (e.g., GE Healthcare or equivalent) (*see* **Note 2**).
2. 33% Percoll solution: Dilute Percoll solution with MCP buffer. Adjust pH to 6.0 and keep in 4 °C for a week.
3. Silicone oil AR200.
4. 400 μL microcentrifuge tube: Snap-cap type.
5. Benchtop centrifuge with rotor for 400 μL microcentrifuge tube.
6. Bathing solution: 500 mM glycine betaine monohydrate, 10 mM $CaCl_2$, 10 mM MES. Keep at room temperature after autoclaving.
7. (R, S)-^{3} H-ABA.
8. Temperature control incubator with light source.
9. Trimmed 200 μL pipette tips: 4–5 mm cut off from narrow end.
10. Liquid nitrogen and storage tank.

11. Scintillation cocktail (e.g., PerkinElmer Ultima Gold or equivalent).
12. 4 mL MicroBeta vial or equivalent scintillation vial.
13. Liquid Scintillation counters.

2.2 ABA Transport Assay with BY-2 Cells

Prepare all solutions using ultrapure water (prepared by purifying deionized water, to attain a sensitivity of 18 MΩ-cm at 25 °C) and analytical grade reagents.

Sterilize all equipment and solutions in a proper way (autoclaving, filtering, UV irradiation) to prevent fungal and/or microbial infection of BY-2 cells.

2.2.1 Culture of BY-2 Suspension

1. Calli of wild-type and/or transformed BY-2 cells (available from various laboratories) grown on solid LS medium in petri dish.
2. BY-2 LS (Linsmaier and Skoog) media: MS salt 4.4 g, myoinositol 0.1 g, KH_2PO_4 0.37 mg, 2,4-dichlorophenoxyacetic acid 0.181 mg, thiamine-hydrochloride 1 mg for 1 L with 3% sucrose and 0.3% Gelrite (only solid media). Adjust pH to 5.7 with 0.1 M KOH and autoclave (*see* **Note 3**).
3. Sterile bench with UV lamp (Laminar Flow Hood).
4. Suitable antibiotics after 0.45 μm filtering, for transformed BY-2 cells culture only.
5. 50 mL conical flasks containing 20 mL of liquid BY-2 LS medium, sealed with aluminum foil: Keep at room temperature after autoclaving.
6. 200 mL conical flasks containing 50 mL of liquid BY-2 LS medium, sealed with aluminum foil: Keep at room temperature after autoclaving.
7. 25 °C Shaking incubator, dark, 120 rpm.
8. Sterile disposable inoculating loops: size 10 μL.
9. Trimmed 1 mL pipette tips: 4–5 mm cut off from narrow end (*see* **Note 4**).
10. Optical Microscope.

2.2.2 ABA Uptake Assay

1. Benchtop swing-bucket centrifuge.
2. Sterile 50 mL falcon tube.
3. Uptake buffer: 20 mM MES, 40 mM Sucrose, 0.5 mM $CaSO_4$, 0.5 mM $CaCl_2$. Adjust pH to 5.7 with 0.1 M KOH. Keep at room temperature after autoclaving.
4. 25 °C Shaking incubator, dark, 120 rpm.
5. Cuvette and spectrophotometer.
6. 200 mL conical flasks sealed with aluminum foil, sterile.

7. (R, S)-^{3} H-ABA.
8. 2 mL Eppendorf tube, round bottom.
9. Mini dry block heater with 2 mL round-bottom holder.
10. Membrane filter: Hydrophilic, pore size 0.45 μm, diameter 25 mm.
11. Fine precise forceps.
12. Millipore 1225 sampling manifold (or equivalent) and vacuum pump.
13. Repetitive dispensing pipette and 50 mL tips (*see* **Note 5**).
14. Scintillation cocktail.
15. 4 mL MicroBeta vial or equivalent scintillation vial.
16. Liquid Scintillation counters.

3 Method

3.1 ABA Transport Assay with Arabidopsis Protoplast

3.1.1 Mesophyll Protoplast Isolation

1. Grow Arabidopsis plants on autoclaved soil in environment-controlled chamber with a short photoperiod (8 h light at 22 °C–16 h dark at 20 °C) under 80–90 μmol m^{-2} s^{-1} and 40–65% relative humidity for 7 weeks.
2. Choose well-expanded rosette leaves from 7–8-week-old plants before flowering (*see* **Note 6**).
3. Abrade the abaxial epidermises with sandpaper (*see* **Note** 7).
4. Immediately float scratched leaves on MCP buffer in petri dish, abaxial side down (*see* **Note 8**).
5. Aspirate MCP buffer well and add 50 mL of digestion buffer to each plate gently (*see* **Note 9**).
6. Incubate in the dark at room temperature for an hour (maximum 2 h) with gentle agitation, less than 50 rpm (*see* **Note 10**).
7. Filter the digestion buffer containing protoplasts with the 100-μm nylon mesh to remove undigested leaf tissues (*see* **Note 11**).
8. Add 500 μL of 1 M $CaCl_2$ gently.
9. Transfer solution into 50 mL round-bottom tube.
10. Centrifuge at 100 × *g* for 1–2 min with swing-bucket rotor.
11. Remove as much supernatant as possible with suction. Fill MCP buffer up to 5 mL and resuspend the protoplast pellet by gentle swirling.
12. Load resuspended protoplast on 5 mL of 21% sucrose solution in 14 mL round-bottom tube (*see* **Note 12**).
13. Centrifuge at 80 × *g* for 5 min with swing-bucket rotor.

14. Aspirate packed health protoplast located in the borderline of gradient solution with pipet several times.
15. Resuspend protoplasts at 1.5×10^6 cells/mL in MCP buffer after counting cells under the microscope using a hemocytometer. Rest the protoplasts by keeping on ice before next experiment (*see* **Note 13**).

3.1.2 ABA Accumulation/Uptake Assay

Here, I do not describe the specific protocol for radioactive tracer usage and disposal of radioactive contaminant. However, strictly comply *Radioactive Safety Guide* in each institute or laboratory whole experimental procedure. Use separated equipment and stuff for isotope only.

1. Build up the 50 μL of 33% Percoll solution and 200 μL of silicone oil gradient in 400 μL microcentrifuge tube (*see* **Note 14** and Fig. 1a).
2. Resuspend the protoplast in 1× volume of a bathing solution (*see* **Note 15**).
3. Incubate protoplast in 25 °C incubator under light for 10–30 min before uptake assay (*see* **Note 16**).
4. Prepare radioactive ABA tracer mix in 1× volume of a bathing solution (*see* **Note 17**).
5. The uptake reaction is started by adding "prepared ABA tracer (*see* **step 4**)" into "protoplast in bathing solution (*see* **step 3**)." Subsequently, take 100 μL of the sample and load on 200 μL of silicone oil in 400 μL microcentrifuge tube after each time interval (*see* **Note 18** and Fig. 1b).
6. Immediately centrifuge at 8000 × *g* for 20 s. Intact protoplasts were collected on 33% Percoll solution after centrifugation through silicone oil gradients (*see* Fig. 1c).
7. Soak the 400 μL microcentrifuge tube in liquid nitrogen for freezing.
8. Cut out cluster of intact protoplast from frozen 400 μL microcentrifuge tube and mix with 3 mL of scintillation cocktail in 4 mL vial (*see* Fig. 1c).
9. Determine the ^{3}H-radioactivity of ABA tracer by liquid scintillation counter.

3.1.3 ABA Secretion/Efflux Assay

Here, I do not describe the specific protocol for radioactive tracer usage and disposal of radioactive contaminant. However, strictly comply *Radioactive Safety Guide* in each institute or laboratory whole experimental procedure. Use separated equipment and stuff for isotope only.

To figure out protoplast incubation time for efflux assay (below **step 5**), ABA uptake assay should be preceded. **Steps 1–4** and **9–12**

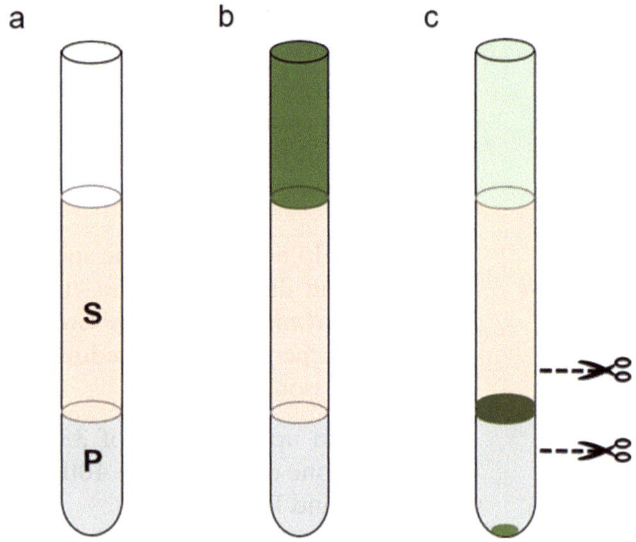

Fig. 1 Percoll–silicone oil gradient usage for intact protoplasts collection after transporting assay. (**a**) Build up the 50 μL of 33% Percoll solution (P) and 200 μL of silicone oil (S) gradient in 400 μL microcentrifuge tube. (**b**) Load 100 μL of the sample (dark green phase) on 200 μL of silicone oil in 400 μL microcentrifuge tube. (**c**) Protoplast separation after centrifuge at 8000 × *g* for 20 s; Upper phase (light green)—light protoplast and broken protoplast, middle phase (dark green)—intact protoplast, low phase (green)—debris. Scissor symbols mark the cutting point after freezing (*see* **step 8**, Subheading 3.1.2)

of Subheading 3.1.3 are the same with **steps 1–4** and **6–9** in Subheading 3.1.2. For more information, please refer to Subheading 3.1.2 and linked note.

1. Build up the 50 μL of 33% Percoll solution and 200 μL of silicone oil gradient in 400 μL microcentrifuge tube.
2. Resuspend the protoplast in 1× volume of a bathing solution.
3. Incubate protoplast in 25 °C incubator under light for 10–30 min before uptake assay.
4. Prepare radioactive ABA tracer mix in 1× volume of a bathing solution.
5. Add radioactive ABA tracer mix in 1× volume of a bathing solution into "prepared protoplast (*see* **step 3**)." Incubate samples in 25 °C incubator under light (*see* **Note 19**).
6. Centrifuge sample solution for 5 min at 50 × *g* with swing-bucket rotor.
7. Discard the supernatant completely.
8. Resuspend the cells with an ABA-free bathing solution and incubate in 25 °C incubator under light for 0, 2, 4, 6, or 10 min. The efflux reaction was started by adding ABA-free

bathing solution. Subsequently, take 100 μL of the sample and load on 200 μL of silicone oil in 400 μL microcentrifuge tube after each time interval.

9. Immediately centrifuge at 8000 × *g* for 20 s. Intact protoplasts are collected on 33% Percoll solution after centrifugation through silicone oil gradients.
10. Soak the 400 μL microcentrifuge tube in liquid nitrogen for freezing.
11. Cut out cluster of intact protoplast from frozen 400 μL microcentrifuge tube and mix with 3 mL of scintillation cocktail in 4 mL scintillation vial.
12. Determine the remained ^{3}H-radioactivity of ABA tracer by liquid scintillation counter.

3.2 ABA Transport Assay with BY-2 Cells

3.2.1 Culture of BY-2 Suspension

1. Prepare calli of transformed BY-2 cells grown on solid LS medium in petri dish (*see* **Note 20**).
2. Rack up BY-2 callus from surface of individual calli with sterile inoculating loops (*see* **Note 21**).
3. Transfer racked up callus to the 50 mL flask with 20 mL LS medium containing antibiotics (*see* **Note 22**).
4. Break up callus with pipetting the culturing medium up and down, slowly.
5. Reseal the flask with a sterile aluminum foil and place flask in a shaking incubator set at 120 rpm, dark, and 25 °C (*see* **Note 23**).
6. Put 7-day cultured flask in sterile bench for 10 min. Collect sunken BY-2 cells using a trimmed 1 mL pipette tip. Transfer collected callus into the 200 mL flask with 50 mL of LS medium containing antibiotics.
7. Subculture; take 10 mL of cells from cultured flask and transfer into new 200 mL flask with 50 mL of LS medium containing antibiotics using pipette aid (*see* **Note 24**).
8. Select synchronized BY-2 cell lines between genotypes after 4–5 times subculture by microscopic observation (*see* **Note 25**).

3.2.2 ABA Uptake Assay

Here, I do not describe the specific protocol for radioactive tracer usage and disposal of radioactive contaminant. However, strictly comply *Radioactive Safety Guide* in each institute or laboratory whole experimental procedure. Use separated equipment and stuff for isotope only.

1. Harvest stationary phase BY-2 cells after 4 days last subculture. Transfer cultured cells into 50 mL falcon tube. Centrifuge at 500 × *g*, 5 min, at 25 °C using swing-bucket rotor.

2. Remove the media and resuspend BY-2 cells with 40 mL of uptake buffer.
3. Transfer resuspended BY-2 cells into 200 mL flask and incubate for 45 min at 120 rpm, dark and 25 °C.
4. Transfer BY-2 cells into 50 mL falcon tube and centrifuge at 500 × *g*, 5 min, at 25 °C using swing-bucket rotor.
5. Resuspend BY-2 cells in uptake buffer at an $OD_{600} = 2$.
6. Transfer resuspended BY-2 cells into 200 mL flask and incubate for 90 min at 120 rpm, dark and 25 °C.
7. Prepare uptake buffer containing 100-fold radiolabeled ABA tracer (*see* **Note 26**).
8. Aliquot cells each 2 mL into 2 mL Eppendorf tube and incubate in mini dry block heater at room temperature with shaking (*see* **Note 27**).
9. ABA uptake reaction is initiated by adding 20 μL of uptake buffer containing 100-fold radiolabeled ABA tracer.
10. Take 0.5 mL of reaction mixture and immediately load on a prewetted membrane filter in Millipore 1225 sampling manifold. Perform three or four technical replicates (0.5 mL each) for each condition or time point and repeated for each genotype of BY-2 cells (*see* **Note 28**).
11. Wash the loaded cells with 2 mL of ice-cold uptake buffer three times, rapidly.
12. Dissolve the filter-bound BY-2 cells with 3 mL of scintillation cocktail in 4 mL scintillation vial. Put the filter with the cells directly into a vial containing scintillation cocktail and vortexing vigorously for 30 s or more.
13. Determine the radioactivity of accumulated ABA tracer into BY-2 cells by liquid scintillation counter.

4 Notes

1. Dissolve enzymes and BSA with stirrer and incubate for 7 min at 55–60 °C for completely dissolution. Incubation for a long time can cause reduction of enzyme activity. Keep in 4 °C after filtering with 0.45 μm filter. Use up enzyme dissolved digestion buffer on the day.
2. Add powdery form of the chemicals in 100% Percoll solution. Dissolve chemicals in 42 °C water bath after vortexing. Keep in 4 °C for a week.
3. Prepare stock solution: Dissolve 37 mg of KH_2PO_4 in 10 mL. Take 100 μL for 1 L. Dissolve 100 mg of Thiamine-hydrochloride in 10 mL. Take 100 μL for 1 L. Use

1000 ppm solution of 2,4-Dichlorophenoxyacetic acid. Take 181 μL for 1 L.

4. Cut off the narrow ends of normal tips with sterile razor blade and autoclave as normal.
5. I use Multipipette Plus (Eppendorf) and 50 mL Combitips with connecter (Eppendorf) for this purpose.
6. The selection of healthy leaves (unstressed, no anthocyanin accumulation) at the proper developmental stage is a very critical factor in experiments.
7. Scratch abaxial side of leaf 2–3 times. Floating leaves show transparent dark green.
8. The volume of MCP buffer is enough to fill a gap between leaves and petri dish surface. The surface of MCP in petri dish should be fill up compactly by floating leaves. Prepare one to two petri dishes per each genotype.
9. Fill up the digestion buffer from the bottom of petri dish using pipette aid, gently. Remove any bubbles above the liquid surface using injection needle.
10. The enzyme dissolved digestion buffer should turn green after a gentle swirling motion, which indicates the release of protoplasts.
11. The size of Arabidopsis mesophyll protoplasts is approximately 30–50 μm. Prepare prewetted nylon mesh with MCP buffer. Lean the petri dish and flush down attached protoplast to leaf tissues with MCP buffer gently. Transfer the MCP buffer with protoplast into digestion buffer with protoplast after filtering. Be gentle with protoplasts, handle them with regular pipettes but trimmed 1 mL pipette tips.
12. Keep the borderline between sucrose solution and MCP buffer containing protoplasts.
13. Although the protoplasts can be kept on ice for at least 24 h, freshly prepared protoplasts should be used in transport assay. Start within an hour.
14. Load Percoll solution first, snap a tube once, overload silicone oil and quick centrifuge.
15. Sample solution for transport assay consist with MCP with protoplast and twofold bathing solution containing radioactive tracer (v/v).

    Final volume of transport assay sample solution (A),

$$A = (\text{number of time point or number of tested condition} + 1) \times 4 \times 100\ \mu L$$

$$1 \times \text{volume} = A/3\ \mu L$$

16. Light intensity is recommended the similar level of growth condition. Use light reflector for efficient irradiation to the protoplasts located in bottom of tube.
17. Finally, approximately 5×10^5 protoplasts are incubated in transport assay sample solution (A), containing 4.5 nM (R, S)-^{3}H-ABA (7.4 kBq; Amersham Biosciences; 1.63 TBq/mmol). Add the required chemicals depending on purpose of experiment or characteristic of tested transporters; the energy source, competitor of substrate, inhibitor of transporters (*see* ref. 12).
18. Each condition and time point is tested with four replicates. Before taking, protoplasts should exist uniform distribution in test tube. Resuspend the protoplast by gentle swirling time to time.
19. Incubation period is settled through antecedent uptake reaction. Incubate protoplasts for time taken to saturation in uptake reaction.
20. Generate Calli of transformed BY-2 cells according to Cho et al. [20]. Prepare ABA transporter expressing BY-2 cells and empty vector or nonfunctional mutated form of ABA transporter expressing BY-2 cells as a negative control. To avoid contamination all procedure should be done in laminar flow sterile bench. All equipment be sterilized proper way before experiment; autoclave, filtering, and UV irradiation.
21. Use fully proliferated calli, more than 2 cm in diameter. Do not use callus in the center of calli.
22. The flasks should be open for as short a time as possible to avoid contamination. Avoid passing anything (e.g., hands) over the open flasks.
23. Lightly flame neck of flask and inside of aluminum foil before resealing.
24. Suspended BY-2 cells should be subcultured every 7 days.
25. In subculture, each of the BY-2 cells is floating independently or at most only in short chains in a culture medium. Depending on transformed cell lines, it is possible that most of the BY-2 cells are floating long chains or short and swollen cell chains. For transport assay, transformed cell lines should be comprised most of the BY-2 cells floating independently or 2–4 cell connected short chain. When you compare transport activity with more than two lines, all lines should consist of similar chain length and width. Please refer Fig. 1 from Kumagai-Sano et al. 2007 [21]).
26. Finally, resuspend BY-2 cells (at an $OD_{600} = 2$) incubated in uptake buffer containing 4.5 nM (R, S)-^{3}H-ABA (7.4 kBq; Amersham Biosciences; 1.63 TBq/mmol).

Add the required chemicals depending on purpose of experiment, characteristic of tested transporters; the energy source, competitor of substrate, inhibitor of transporters at 100-fold to final concentration with ABA tracer.

27. The shaking frequency should be high enough to prevent BY-2 cell sink down.
28. Incubate membrane filters shortly in uptake buffer for prewetting. 12-filter plate position of Millipore 1225 sampling manifold should be covered with prewetted membrane filter or supplement plug and turn on connected vacuum pump before reaction mixture loading.

Acknowledgments

This work was supported by Brain Pool Program through the NRF funded by the Ministry of Science and ICT (2019H1D3A2A02102582) and Ambizione fellowship of the Swiss National Science Foundation (PZ00P3_168041).

References

1. Finkelstein R (2013) Abscisic acid synthesis and response. Arabidopsis Book 11:e0166
2. Endo A, Sawada Y, Takahashi H, Okamoto M, Ikegami K, Koiwai H, Seo M, Toyomasu T, Mitsuhashi W, Shinozaki K, Nakazono M, Kamiya Y, Koshiba T, Nambara E (2008) Drought induction of Arabidopsis 9-cis-epoxycarotenoid dioxygenase occurs in vascular parenchyma cells. Plant Physiol 147: 1984–1993
3. Kuromori T, Sugimoto E, Shinozaki K (2014) Intertissue signal transfer of abscisic acid from vascular cells to guard cells. Plant Physiol 164: 1587–1592
4. Christmann A, Weiler EW, Steudle E, Grill E (2007) A hydraulic signal in root-to-shoot signalling of water shortage. Plant J 52:167–174
5. Ma Y, Szostkiewicz I, Korte A, Moes D, Yang Y, Christmann A, Grill E (2009) Regulators of PP2C phosphatase activity function as abscisic acid sensors. Science 324:1064–1068
6. Park SY, Fung P, Nishimura N, Jensen DR, Fujii H, Zhao Y, Lumba S, Santiago J, Rodrigues A, Chow TF, Alfred SE, Bonetta D, Finkelstein R, Provart NJ, Desveaux D, Rodriguez PL, McCourt P, Zhu JK, Schroeder JI, Volkman BF, Cutler SR (2009) Abscisic acid inhibits type 2C protein phosphatases via the PYR/PYL family of START proteins. Science 324:1068–1071
7. Gonzalez-Guzman M, Pizzio GA, Antoni R, Vera-Sirera F, Merilo F, Bassel GW, Fernández MA, Holdsworth MJ, Perez-Amador MA, Kollist H, Rodriguez PL (2012) Arabidopsis PYR/PYL/RCAR receptors play a major role in quantitative regulation of stomatal aperture and transcriptional response to abscisic acid. Plant Cell 24:2483–2496
8. Daeter W, Hartung W (1993) The permeability of the epidermal cell plasma membrane of barley leaves to abscisic acid. Planta 191:41–47
9. Wilkinson S, Davies WJ (1997) Xylem sap pH increase: a drought signal received at the apoplastic face of the guard cell that involves the suppression of saturable abscisic acid uptake by the epidermal symplast. Plant Physiol 113: 559–573
10. Kang J, Lee Y, Martinoia E (2020) How can we interpret the large number and diversity of aba transporters? Prog Bot 82:233–257
11. Kuromori T, Miyaji T, Yabuuchi H, Shimizu H, Sugimoto E, Kamiya A, Moriyama Y, Shinozaki K (2010) ABC transporter AtABCG25 is involved in abscisic acid transport and responses. Proc Natl Acad Sci U S A 107:2361–2366
12. Kang J, Hwang JU, Lee M, Kim YY, Assmann SM, Martinoia E, Lee Y (2010) PDR-type ABC transporter mediates cellular uptake of

the phytohormone abscisic acid. Proc Natl Acad Sci U S A 107:2355–2360

13. Kanno Y, Hanada A, Chiba Y, Ichikawa T, Nakazawa M, Matsui M, Koshiba T, Kamiya Y, Seo M (2012) Identification of an abscisic acid transporter by functional screening using the receptor complex as a sensor. Proc Natl Acad Sci U S A 109:9653–9658
14. Léran S, Noguero M, Corratgé-Faillie C, Boursiac Y, Brachet C, Lacombe B (2020) Functional characterization of the *Arabidopsis* abscisic acid transporters NPF4.5 and NPF4.6 in *Xenopus* oocytes. Front Plant Sci 11:144
15. Zhang H, Zhu H, Pan Y, Yu Y, Luan S, Li L (2014) A DTX/MATE-type transporter facilitates abscisic acid efflux and modulates ABA sensitivity and drought tolerance in Arabidopsis. Mol Plant 7:1522–1532
16. Krattinger SG, Kang J, Bräunlich S, Boni R, Chauhan H, Selter LL, Robinson MD, Schmid MW, Wiederhold E, Hensel G, Kumlehn J, Sucher J, Martinoia E, Keller B (2019) Abscisic acid is a substrate of the ABC transporter encoded by the durable wheat disease resistance gene Lr34. New Phytol 223:853–866
17. Yao L, Cheng X, Gu Z, Huang W, Li S, Wang L, Wang Y-F, Xu P, Ma H, Ge X (2018) The AWPM-19 family protein OsPM1 mediates abscisic acid influx and drought response in Rice. Plant Cell 30:1258–1276
18. Pawela A, Banasiak J, Biała W, Martinoia E, Jasiński M (2019) MtABCG20 is an ABA exporter influencing root morphology and seed germination of *Medicago truncatula*. Plant J 98:511–523
19. Kuromori T, Sugimoto E, Shinozaki K (2021) *Brachypodium* BdABCG25 is a homolog of Arabidopsis AtABCG25 involved in the transport of abscisic acid. FEBS Lett 595:954–959. https://doi.org/10.1002/1873-3468.13925
20. Cho M, Lee SH, Cho HT (2007) P-glycoprotein4 displays auxin efflux transporter-like action in Arabidopsis root hair cells and tobacco cells. Plant Cell 19: 3930–3943
21. Kumagai-Sano F, Hayashi T, Sano T, Hasezawa S (2006) Cell cycle synchronization of tobacco BY-2 cells. Nat Protoc 1:2621–2627

Chapter 7

Screening of ABA Transporters by a Yeast Two-Hybrid System-Based Screening Using the Receptor Complex as a Sensor

Shunsuke Watanabe, Yuri Kanno, and Mitsunori Seo

Abstract

Abscisic acid (ABA) is a plant hormone that is involved in many physiological events and is present in most plant tissues. How ABA moves within plants has not been well understood. To characterize the physiological consequences as well as the underlying molecular events responsible for ABA movement, it is crucial to identify the transporter proteins. Several approaches have been successful in identifying ABA transporters. In this chapter, we outline a methodology to directly identify proteins capable of transporting ABA in a heterologous yeast system. In our assay, the ABA receptor [PYRABACTIN RESISTANCE1 (PYR1) and PYR1-LIKE (PYL), also known as REGULATORY COMPONENTS OF ABA RECEPTOR (RCAR)] and the protein phosphatases of type 2C (PP2C) coreceptor interact in an ABA-dependent manner. A yeast two-hybrid (Y2H) system is used to monitor interactions between the receptor and PP2C and, hence, the ABA concentration in the yeast cells. Screening cDNAs that induce or inhibit the receptor–PP2C interaction in the presence of ABA allows us to identify candidate transporters. ABA transport activities of the putative transporter proteins can be validated by quantifying hormone levels in the yeast cells using liquid chromatography tandem-mass spectrometry (LC-MS/MS).

Key words Plant hormones, Abscisic acid, Transporters, Yeast two-hybrid system, LC-MS/MS

1 Introduction

Abscisic acid (ABA) is a plant hormone that plays crucial roles in various aspects of plant growth and development such as seed formation and germination, and plant responses to abiotic and biotic stresses [1–3]. A number of genes involved in the biosynthesis, catabolism, perception, and signal transduction of ABA have been identified, especially in the model plant Arabidopsis [4, 5]. Spatial expression patterns and protein localization of ABA biosynthetic genes/enzymes indicate that ABA is actively synthesized in leaf vascular tissues during water stress responses [6]. These observations imply that ABA synthesized in vascular tissues is

Takuya Yoshida (ed.), *Abscisic Acid: Methods and Protocols*,
Methods in Molecular Biology, vol. 2462, https://doi.org/10.1007/978-1-0716-2156-1_7,

translocated to guard cells to induce stomatal closure. Furthermore, ABA responses take place not only in guard cells but also in other cell and tissue types [6]. Thus, ABA concentrations at the multiple target sites must be regulated precisely through active transport mechanisms.

The movement of ABA within plants is regulated in principle by combinations of sequential ABA export and import events from and into cells mediated by transmembrane transporters. Despite significant efforts to identify factors involved in ABA-mediated physiological responses, ABA transporters were unknown until a decade ago possibly due to functional redundancies. Several approaches contributed to overcoming the problem, and three types of ABA transporters, namely, the ATP-binding cassette (ABC), the NITRATE TRANSPORTER 1/PEPTIDE TRANS PORTER FAMILY (NPF), and detoxification efflux carrier (DTX) [also known as multidrug and toxic compound extrusion (MATE)] proteins, have been identified so far [6–13]. Here, we describe one of the approaches that can be used to comprehensively look for potential ABA transporters in a less biased manner even in nonmodel plant species.

The ABA receptors PYRABACTIN RESISTANCE1 (PYR1) and PYR1-LIKE (PYL) proteins [also known as REGULATORY C OMPONENTS OF ABA RECEPTORs (RCARs)] bind to protein phosphatases of type 2C (PP2C) upon recognition of the ligand, releasing SNF1-related protein kinase 2 s (SnRK2s) from inhibition by PP2Cs [14, 15]. In turn, the kinases activate downstream ABA signaling pathways through phosphorylation [16]. Our idea was to utilize the ABA receptor and PP2C to detect ABA transport activities of proteins expressed in yeast cells. ABA-dependent interactions between the receptor and PP2C can be monitored by the yeast two-hybrid (Y2H) system. Yeast cells containing a Y2H system with the ABA receptor and PP2C cannot grow on normal selection media, but they can grow on media if ABA is exogenously applied [14–16]. What we expected was if an ABA importer was expressed in the system, the yeast cells would be able to grow on selection media in response to much lower concentrations of ABA compared to cells not producing the transporter (Fig. 1). In contrast, expression of an ABA exporter in the system would inhibit yeast growth on selection media containing ABA. This system enabled us to conduct a large-scale screening for potential ABA transporters from cDNA libraries cloned in a yeast expression vector. ABA transport activities of the candidate transporters can be validated by directly analyzing the compounds taken into the yeast cells using liquid chromatography tandem-mass spectrometry (LC-MS/MS).

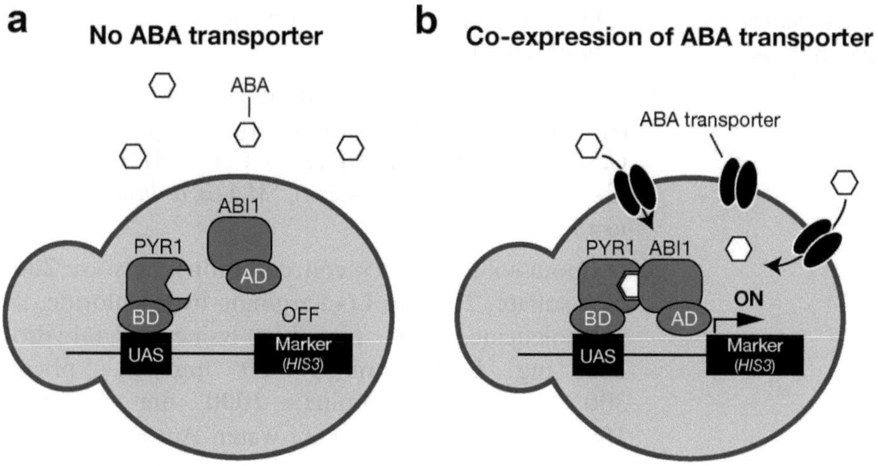

Fig. 1 Detection of ABA transport activities using a yeast two-hybrid (Y2H) system with the ABA receptor complex. ABA-dependent interactions between PYR1 and ABI1 can be reconstituted in yeast cells. The GAL4 DNA-binding domain (BD) fused to PYR1 binds to the upstream activating sequence (UAS) of a marker gene (*HIS3* in our system). If PYR1 interacts with ABI1, the GAL4-activation domain (AD) promotes transcription of the marker gene. (**a**) ABA is not actively taken into yeast cells in the absence of specific transporters. Therefore, PYR1-ABI1 interactions are not induced when ABA concentrations in the media are relatively low and the selectable marker is not transcribed. (**b**) If an ABA transporter (importer) is expressed by the yeast cells, the cellular ABA content increases even under low ABA concentrations. The expression of the ABA transporter (importer) induces PYR1-ABI1 interactions, and the marker is actively transcribed

2 Materials

Prepare all solutions at room temperature using ultrapure water (e.g., Milli-Q water with ≥18 MW/cm resistance at 25 °C), unless otherwise noted. All chemicals and solvents should be of analytical or higher grades.

2.1 Vector Construction

1. Template cDNAs and primers to amplify coding sequences (CDSs) of an ABA receptor and PP2C (*see* **Note 1**).
2. Cloning (entry) vector (e.g., Invitrogen pENTR/D-TOPO or equivalent) (*see* **Note 2**).
3. Bait and pray vectors (e.g., Invitrogen pDEST22 [*TRP1* marker] and pDEST32 [*LEU2* marker] or equivalent) (*see* **Note 3**).
4. A yeast expression vector (e.g., pYES-DEST52-*ADH1pro* [pYES-DEST52 in which *GAL1* promoter has been replaced with *ADH1* promoter] [*URA3* marker] or equivalent) (*see* **Notes 4** and **5**).

2.2 Transformation and Culture of Yeast Cells

1. cDNA Library Construction Kit (e.g., Thermo Fisher CloneMiner II).
2. *Saccharomyces cerevisiae* [strain PJ69-4A (for Y2H): *MATa trp1-901 leu2-3112 ura3-52 his3-200 gal4Δ gal80Δ Met2::GAL7-lacZ LYS2::GAL1-HIS3 GAL2-ADE2*; strain INVSc1 (for direct transport assays): *MATa/α his3Δ1 leu2 trp1-289 ura3-52*].
3. Dropout solution (10× concentration): Dissolve 200 mg adenine sulfate, 200 mg L(+)-arginine hydrochloride, 300 mg L(+)-isoleucine, 300 mg L(+)-lysine monohydrochloride, 200 mg L-methionine, 500 mg L(−)-phenylalanine, 300 mg L(−)-threonine, 2000 mg L-tyrosine, and 1500 mg L-valine in 1 L of water. Autoclave at 120 °C for 15 min. The solution can be stored at 4 °C for up to 6 months.
4. Histidine (H) solution (100× concentration): Dissolve 2 g L-histidine hydrochloride monohydrate in 1 L of water and sterilize with a 0.45 μm pore filter.
5. Leucine (L) solution (100× concentration): Dissolve 10 g L-leucine in 1 L of water and sterilize with a 0.45 μm pore filter.
6. Tryptophan (W) solution (100× concentration): Dissolve 2 g L-tryptophan in 1 L of water and sterilize with a 0.45 μm pore filter.
7. Uracil (U) solution (100× concentration): Dissolve 2 g uracil in 1 L of water and sterilize with a 0.45 μm pore filter.
8. 20% (v/v) D(+)-glucose solution: Sterilize with a 0.45 μm pore filter.
9. 20% (v/v) D(+)-galactose solution: Sterilize with a 0.45 μm pore filter (*see* **Note 6**).
10. 20% (v/v) D(+)-raffinose solution: Sterilize with a 0.45 μm pore filter.
11. Synthetic defined (SD) media: Dissolve 6.7 g yeast nitrogen base without amino acids in 700 mL of water. Autoclave at 120 °C for 15 min. After cooling the media, add 100 mL of the dropout solution and 10 mL of the H, L, W, and U solutions. As a carbon source, add 100 mL of the glucose solution. Fill up to 1 L with sterile water. Add 2.0% (v/v) agar before autoclave for solid media.
12. SD-LW: Dissolve 6.7 g yeast nitrogen base without amino acids in 700 mL of water. Autoclave at 120 °C for 15 min. After cooling the media, add 100 mL of the dropout solution and 10 mL of the H and U solutions. As a carbon source, add 100 mL of the glucose solution. Fill up to 1 L with sterile water. Add 2.0% (v/v) agar before autoclave for solid media.

13. SD-LWH: Dissolve 6.7 g yeast nitrogen base without amino acids in 700 mL of water. Autoclave at 120 °C for 15 min. After cooling the media, add 100 mL of the dropout solution and 10 mL of the U solution. As a carbon source, add 100 mL of the glucose solution. Fill up to 1 L with sterile water. Add 2.0% (v/v) agar before autoclave for solid media.
14. SD-LWU: Dissolve 6.7 g yeast nitrogen base without amino acids in 700 mL of water. Autoclave at 120 °C for 15 min. After cooling the media, add 100 mL of the dropout solution and 10 mL of the H solution. As a carbon source, add 100 mL of the glucose solution. Fill up to 1 L with sterile water. Add 2.0% (v/v) agar before autoclave for solid media.
15. SD-LWUH: Dissolve 6.7 g yeast nitrogen base without amino acids in 700 mL of water. Autoclave at 120 °C for 15 min. After cooling the media, add 100 mL of the dropout solution. As a carbon source, add 100 mL of the glucose solution. Fill up to 1 L with sterile water. Add 2.0% (v/v) agar before autoclave for solid media.
16. Synthetic galactose (SG) media: Dissolve 6.7 g yeast nitrogen base without amino acids in 700 mL of water. Autoclave at 120 °C for 15 min. After cooling the media, add 100 mL of the dropout solution and 10 mL of the H, L, W, and U solutions. As a carbon source, add 100 mL of the galactose solution and 50 mL of the raffinose solution. Fill up to 1 L with sterile water. Add 2.0% (v/v) agar before autoclave for solid media.
17. SG-U: Dissolve 6.7 g yeast Nitrogen Base without Amino Acids in 700 mL of water. Autoclave at 120 °C for 15 min. After cooling the media, add 100 mL of the dropout solution and 10 mL of the H, L, and W solutions. As a carbon source, add 100 mL of the galactose solution and 50 mL of the raffinose solution. Fill up to 1 L with sterile water. Add 2.0% (v/v) agar before autoclave for solid media.
18. Tris–EDTA (TE) buffer (pH 7.5): 10 mM tris (hydroxymethyl)aminomethane, 1 mM ethylenediaminetetraacetic acid. Adjust pH to 7.5 with 5 M hydrochloric acid and autoclave at 120 °C for 15 min.
19. ABA solution (1000× concentration): Dissolve in DMSO at desired concentrations and add to the media after autoclave.
20. 1 M 3-amino-1,2,4-triazole (3-AT) solution: Sterilize with a 0.45 μm pore filter and add to the media after autoclave.
21. Yeast plasmid isolation kit.
22. 1.5 mL centrifugation tubes.
23. 15 mL centrifugation tubes.
24. 50 mL centrifugation tubes.

25. 500 mL baffled flasks.
26. Square petri dishes (10 cm × 14 cm).

2.3 Direct Transport Assay of ABA Using Yeast Cells

1. 1 M potassium dihydrogen phosphate (KH_2PO_4) solution: Autoclave at 120 °C for 15 min.
2. 1 M dipotassium hydrogen phosphate (K_2HPO_4) solution: Autoclave at 120 °C for 15 min.
3. 50 mM potassium phosphate buffer (KPB), pH 5.8: Mix 46 mL of the 1 M KH_2PO_4 solution and 4 mL of the 1 M K_2HPO_4 solution and fill up to 1 L with water.
4. 50 mM KPB, pH 7.0: Mix 19.2 mL of the 1 M KH_2PO_4 solution and 30.8 mL of the 1 M K_2HPO_4 solution and fill up to 1 L with water.
5. ABA solution (1000× concentration): Dissolve in DMSO at desired concentrations and add to the media after autoclave.

2.4 Extraction and Purification of ABA from Yeast Cells

1. Reverse phase weak anion exchange column, 1 cc capacity (e.g., Waters Oasis WAX or equivalent).
2. Nitrogen evaporator.
3. 10 pg/μL isotope-labeled ABA (e.g., [$^2H_6$]ABA) dissolved in 50% (v/v) methanol.
4. 80% (v/v) acetone containing 1% (v/v) acetic acid.
5. 1% (v/v) acetic acid.
6. Absolute acetonitrile.
7. 0.1 M potassium hydroxide (KOH).
8. 80% (v/v) acetonitrile.
9. 80% (v/v) acetonitrile containing 1% (v/v) acetic acid.

2.5 LC-MS/MS Analysis

1. LC-MS/MS system.
2. C18 column (e.g., Waters ACQUITY UPLC BEH C18 column or equivalent).
3. Distilled water (LC/MS grade) containing 0.01% (v/v) acetic acid.
4. Acetonitrile (LC/MS grade) containing 0.05% acetic acid.

3 Methods

3.1 Construction of Y2H-Based ABA Sensors

1. Amplify CDSs of an ABA receptor (*PYR1*) and a PP2C (*ABI1*) from Arabidopsis cDNAs by PCR using appropriate primer combinations and clone them into a cloning (entry) vector (e.g., pENTR/D-TOPO).

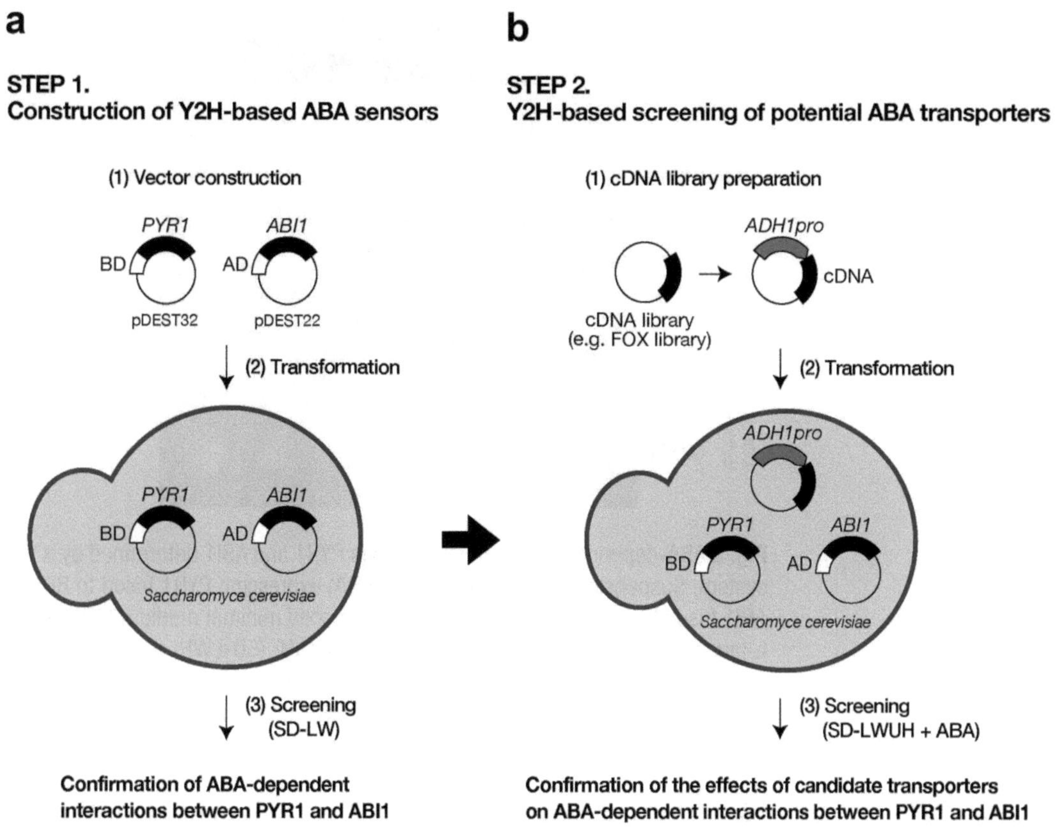

Fig. 2 Construction of a Y2H-based ABA sensor and screening of ABA transporters using the system. (**a**) (STEP 1) PYR1 and ABI1 are cloned into Y2H prey (pDEST32) and bait (pDEST22) vectors, respectively. Both constructs are introduced into a single yeast strain (PJ69-4A). Transformants are selected on standard minimal medium (SD) lacking L-leucine (L), L-tryptophan (W) (SD-LW); their ABA-dependent growth on SD lacking L, W and L-histidine (H) (SD-LWH) is confirmed (Fig. 3). (**b**) (STEP 2) To screen for ABA transporters, cDNA libraries are constructed for a yeast expression vector in which the *GAL1* promoter has been replaced with the *ADH1* promoter (pYES-DEST52-*ADH1pro*). Yeast containing the Y2H system obtained in STEP 1 is transformed with the cDNA libraries cloned in pYES-DEST52-*ADH1pro*. Colonies that can grow on SD lacking L, W, Uracil (U) and H (SD-LWUH) and supplemented with ABA at concentrations for which PYR1-ABI1 interactions are not induced in the absence of transporters are screened

2. After confirming the sequences, transfer the *AtPYR1* CDS into a bait vector (pDEST32) to express proteins fused to the GAL4 DNA-binding domain (BD) and transfer the *AtABI1* CDS into a prey vector (pDEST22) to express proteins fused to the GAL4-activation domain (AD) [Fig. 2a(1)] (*see* **Note 7**).
3. Transform yeast PJ69-4A cells with pDEST32 containing *AtPYR1* (*AtPYR1*/pDEST32) and pDEST22 containing *AtABI1* (*AtABI1*/pDEST22) [Fig. 2a(2)].
4. Select transformants on standard solid SD media lacking L and W (SD-LW) at 30 °C for 3 days [Fig. 2a(3)].

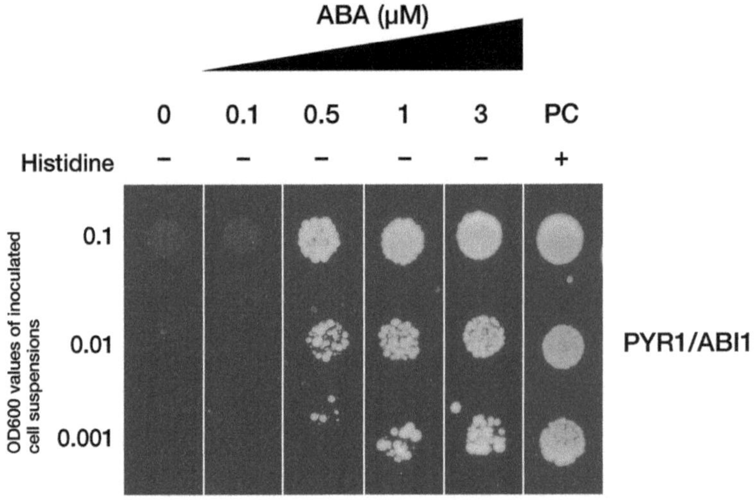

Fig. 3 ABA-dependent interactions between PYR1 and ABI1 determined by a Y2H system. Suspensions of yeast cells (PJ69-4A) expressing PYR1 fused to BD and ABI1 fused to AD were spotted on a standard solid minimal medium (SD) lacking L-leucine (L), L-tryptophan (W), and L-histidine (H) (SD-LWH) that was supplemented with 1 mM 3-AT and selected concentrations of ABA. As a positive control (PC), histidine was added to SD-LWH (SD-LW). Photos were taken 3 days after incubation at 30 °C

5. Culture the transformants in 2 mL of liquid SD-LW media overnight at 30 °C.
6. Dilute the yeast cells with TE buffer and prepare a dilution series (OD600 = 0.1, 0.01 and 0.001).
7. Drop 10 μL of the cell suspensions onto solid SD-LWH media containing 1 mM 3-AT and a range of concentrations of (±)-ABA (*see* **Notes 8** and **9**).
8. Observe yeast growth frequently for 3–6 days after incubation at 30 °C to evaluate whether the PYR1-ABI1 interactions (i.e., yeast growth on selection media) are dependent on ABA (Fig. 3) (*see* **Note 10**).

3.2 Y2H-Based Screening of Potential ABA Transporters

1. Prepare cDNA libraries in a yeast expression vector [Fig. 2b(1)] (*see* **Notes 11**).
2. Transform yeast PJ69-4A cells containing *AtPYR1*/pDEST32 and *AtABI1*/pDEST22 constructs with a cDNA library prepared in pYES-DEST52-*ADH1pro* [Fig. 2b(2)] (*see* **Note 12**).
3. Spread the transformants on solid SD lacking L, W, U, and H (SD-LWUH) supplemented with 1 mM 3-AT and 0.1 μM (±)-ABA and incubate at 30 °C for up to 14 days (or more if necessary) to identify potential ABA importers [Fig. 2b(3)] (*see* **Notes 13–15**).

4. Propagate the first positive colonies selected on fresh solid SD-LWU media at 30 °C for approximately 3 days.
5. Dissolve the yeast cells in TE buffer and prepare a dilution series (OD600 = 0.1, 0.01 and 0.001).
6. Drop 10 μL of the cell suspensions onto solid SD-LWUH media containing 1 mM 3-AT and a range of (±)-ABA concentrations.
7. Observe yeast growth frequently for 3–6 days after incubation at 30 °C to evaluate whether the PYR1-ABI1 interactions (i.e., yeast growth on selection media) are dependent on ABA.
8. Extract DNA from the second positive colonies obtained in the above step using for example an yeast plasmid isolation kit.
9. Amplify cDNAs cloned in pYES-DEST52-*ADH1pro* using primers designed based on the vector sequence.
10. Sequence the amplified DNA fragments to identify candidate transporter genes (*see* **Note 16**).

3.3 Validation of ABA Transport Activities of the Candidates by Y2H Assays

1. Amplify CDS regions of the candidate ABA transporter genes obtained in Subheading 3.2 by PCR and clone into pYES-DEST52-*ADH1pro*.
2. Transform yeast PJ69-4A cells containing *AtPYR1*/pDEST32 and *AtABI1*/pDEST22 constructs with one candidate gene cloned in pYES-DEST52-*ADH1pro* constructed in **step 1**.
3. After selection of positive colonies that grow well on SD-LWU plates at 30 °C for 3 days, culture the colonies overnight in 2 mL of liquid SD-LWU media in 15 mL tubes with shaking at 30 °C (*see* **Note 17**).
4. Dilute the cultures with TE buffer to adjust the OD600 = 0.1, 0.01, and 0.001.
5. Drop 10 μL of yeast suspensions prepared onto solid SD-LWUH media containing 1 mM 3-AT and a range of (±)-ABA concentrations.
6. Observe growth of the yeast cells frequently during 3–6 days after incubation at 30 °C to evaluate ABA-dependent PYR1-ABI1 interactions (Fig. 4) (*see* **Note 18**).

3.4 Validation of ABA Transport Activities of the Candidates by Direct Analysis with LC-MS/MS

1. Clone CDSs of the ABA transporter candidate genes whose functions were validated in Subheading 3.3 into an expression vector (e.g., pYES-DEST52), and transform the constructs into yeast INVSc1 cells (*see* **Note 19**).
2. After selection of positive colonies on SD-U plates for 3 days at 30 °C, cultivate the cells overnight in 5 mL of liquid SG-U media in 50 mL tubes with shaking at 30 °C (*see* **Note 17**).

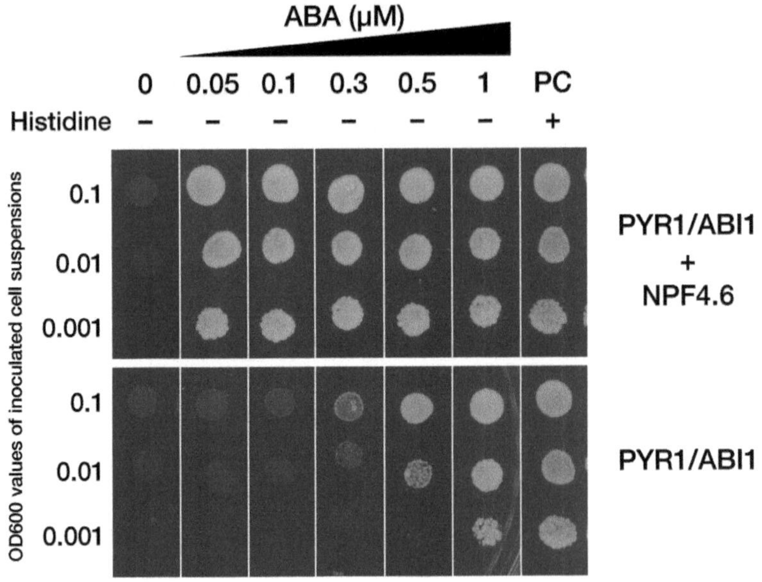

Fig. 4 Effects of an ABA importer (NPF4.6) on the interaction between PYR1 and ABI1 determined by a Y2H system. NPF4.6 was expressed in yeast (PJ69-4A) containing an ABA sensor (the Y2H system with PYR1 and ABI1) and cell suspensions were spotted on SD-LWUH that was supplemented with 1 mM 3-AT and ABA at selected concentrations. As a positive control (PC), histidine was added to SD-LWUH (SD-LWU). Yeast cells containing the ABA sensor (not expressing NPF4.6) were spotted as a comparison. Photos were taken 3 days after incubation at 30 °C

3. Inoculate the overnight preculture into 100 mL of fresh SG-U media in 500 mL baffled flasks and incubate the cell suspensions overnight with shaking at 30 °C.
4. Centrifuge the culture at 3000 × *g* for 5 min and then resuspend the cell pellets in 50 mL of 50 mM KPB (pH 5.8) to wash the cells. Repeat the washing step (three times in total) and resuspend the cell pellets in 50 mM KPB (pH 5.8) to adjust the OD600 = 10.
5. Prepare 50 mM KPB (pH 5.8) containing ABA at twice the concentrations of the final reaction mixtures (*see* **Notes 20–22**). Mix 500 μL of the 2× ABA solution (in 50 mM KPB, pH 5.8) with 500 μL of yeast suspensions (in 50 mM KPB, pH 5.8) prepared in **step 4** in 1.5 mL tubes to initiate the reactions (*see* **Note 23**).
6. Place the tubes at 25 °C for 2–10 min and collect cells by centrifugation at 21,000 × *g* for 1 min (*see* **Note 24**).
7. After discarding the supernatants, resuspend the pellets in 1 mL of 50 mM KPB (pH 7.0) (*see* **Note 25**) and collect cells by centrifugation at 21,000 × *g* for 1 min. Repeat this step three times in total.

8. After discarding the supernatants, freeze the pellets in liquid nitrogen, and store at −80 °C until it is time to extract and purify ABA.
9. Add 1 mL of 80% (v/v) acetone containing 1% (v/v) acetic acid and 50 μL of $[^2H_6]$ABA solution (*see* **Note 26**) into the tubes containing the frozen yeast cells. After vortexing vigorously, place the tubes at 4 °C for 2 h to extract ABA.
10. After centrifuging the extracts at 21,000 × *g* for 5 min, transfer the supernatants to a new tube, and completely dry using a stream of N_2 with a nitrogen evaporator.
11. Dissolve the extracts in 1 mL of 1% (v/v) acetic acid and load the samples on reverse phase weak anion exchange columns (1 cc capacity) that are equilibrated successively with 1 mL of acetonitrile, 1 mL of 0.1 M KOH and 1 mL of 1% (v/v) acetic acid before use.
12. Wash the columns with 1 mL of 1% (v/v) acetic acid and then with 1 mL of 80% (v/v) acetonitrile.
13. Elute fractions containing ABA with 2 mL of 80% (v/v) acetonitrile containing 1% (v/v) acetic acid. Completely dry the eluates under a stream of N_2.
14. Dissolve the extracts in 50 μL of 1% (v/v) acetic acid and quantify ABA levels by LC-MS/MS analysis. Conditions for LC and MS/MS are summarized in Table 1.

Table 1
LC-MS/MS conditions

Instrument	Parameter	Conditions
LC	System	Nexera system (SHIMADZU)
	Column	ACQUITY UPLC BEH C18 column 2.1 × 50 mm (waters)
	Column temperature	40 °C
	Flow rate	400 μL min^{-1}
	Solvent A	Water containing 0.01% (v/v) acetic acid
	Solvent B	Acetonitrile containing 0.05% (v/v) acetic acid
	Gradient (composition of solvent B)	Constant at 3% for 0.5 min
		Linear gradient from 3% to 40% over 7 min
MS/MS	System	Triple TOF 5600 (AB SCIEX)
	Polarity of ESI	Negative
	Ionspray voltage	−3.5 kV
	Desolvation temperature	600 °C
	Declustering potential	−90 V
	Collision energy	−15 V
	Scan range	100–1000 *m/z*
	Precursor ion/qualifier ion	$[^2H_6]$ABA: 269.1/159.1; ABA: 263.1/153.1 *m/z*

4 Notes

1. Interactions between Arabidopsis PYR1 (also known as RCAR11; At4g17870) and ABI1 (At4g26080) have been well characterized and, thus, we normally use this combination. Other receptor-PP2C pairs could be substituted.
2. Other cloning systems can be used, but we are describing a simple method using the Gateway Cloning Technology as an example.
3. Other Y2H systems can be used, but we are describing a simple method using the Gateway Cloning Technology as an example.
4. This vector is used to express candidate ABA transporters in yeast containing a Y2H system with an ABA receptor and PP2C. If necessary, the selectable marker could be replaced with another marker depending on the yeast strain and the Y2H system used for the experiments.
5. For generating a constitutive expression pYES-DEST52-*ADH1pro* vector in yeast cells, replace GAL1 promoter in pYES-DEST52, which is a galactose-inducible promoter, with *ADH1pro* (e.g. amplified from pGBT9 vector) by conventional restriction digestion and ligation methods.
6. Galactose should be of high purity grade [$\leq$0.05% (v/v) glucose at least] to ensure maximum induction of the GAL1 promoter.
7. *PYR1* and *ABI1* CDSs can be cloned in pDEST22 and pDEST32, respectively, but we have not tested if they will work.
8. Both ($\pm$)-ABA (racemic mixture) and (+)-ABA (naturally occurring ABA in plants) are commercially available (https://pubchem.ncbi.nlm.nih.gov/). As ($\pm$)-ABA is cheaper than (+)-ABA, we use the racemic mixture for Y2H assays.
9. Optimize the concentration of 3-AT if necessary.
10. In our experience, ABA-dependent interactions between PYR1 and ABI1 become undetectable 1–2 months after transformation for unknown reasons. Therefore, we transform the constructs each time before a set of experiments.
11. We use pYES-DEST52 of which the *GAL1* promoter has been replaced with the *ADH1* promoter (pYES-DEST52-*ADH1pro*). Yeast strains used for the GAL4-based Y2H system are mutants for *GAL4* and thus the *GAL1* promoter is not active in these strains.
12. High transformation efficiency ($\geq 10^5$ CFU/μg plasmid DNA) is desired for the first screening. For this purpose, use higher purity grade carrier DNA for the lithium acetate transformation.

13. We use 100 μL cell suspension (OD600 ≈ 12) and 1 μg plasmid DNA per square plate.
14. To calculate transformation efficiencies, spread a portion of the transformants on SD-LWU. In total, we screened about 1×10^6 cDNA clones (transformants) in the first screening.
15. ABA exporters can theoretically be isolated by using negative selectable markers, for example, *URA3* in combination with 5-fluoroorotic acid (5-FOA) [17].
16. We expect that not only ABA transporters but also other kinds of proteins that affect the formation of the ABA receptor complex can be isolated by this system. Conducting direct transport assays with the candidate ABA transporters is essential to assess their function.
17. We typically mix approximately 10 independent colonies to normalize variation and repeat the same experiment at least twice.
18. We have not observed positive or negative effects of *AtABCG25* [18] and *AtABCG40* [19], which have been characterized as an ABA exporter and an ABA importer, respectively, on ABA-dependent PYR1-ABI1 interactions. This result is possibly because the proteins are not properly expressed in yeast cells.
19. pYES-DEST52-*ADH1pro* can also be used for protein expression in INVSc1; however, INVSc1 is not a *gal4* mutant and thus the galactose-inducible *GAL1* promoter is active in the strain. As the *GAL1* promoter activity is higher than the *ADH1* promoter activity, we typically use the original pYES-DEST52 vector for the direct transport assays.
20. Some transporters might discriminate between the (+) and (−) enantiomers of ABA. Choose (+)-ABA or (±)-ABA depending on the purpose of your experiments.
21. Prepare 1000× stocks of ABA in DMSO so that the final DMSO content in the reaction mixture will be 0.1%. Adding high concentrations of ABA directly into the reaction mixture varies the results. Thus, it is important to prepare a 2× solution in the reaction buffer (50 mM KPB, pH 5.8) in advance and add the same volume of yeast suspension (prepared in 50 mM KPB, pH 5.8) to initiate the assays.
22. The pH of the reaction mixture will affect the activities of some transporters. For example, proton-dependent transporters such as NPFs show higher activities in lower pH conditions. Also, the membrane permeability of acids, including ABA, increases in lower pH conditions. Some transporters might also discriminate between protonated and ionized forms of

ABA. As the populations of the two forms of ABA change depending on pH, transport activities might also be affected by pH for this reason.

23. We normally test ABA transport activities in the presence of 1 or 10 μM ABA for the first time and change the concentrations, if necessary, depending on the purpose of the experiment.
24. Activities must be determined before the reactions reach a plateau.
25. We define the addition of KPB (pH 7.0) for the first wash as the end of the reaction. For example, for a 5 min reaction, incubate the mixture at 25 °C for 3 min, centrifuge for 1 min, discard the supernatant, and add the washing buffer in 1 min.
26. Isotope-labeled ABA is used as an internal standard of LC-MS/MS analysis to correct purification, ionization, and fragmentation efficiencies in samples. ABA levels are quantified based on the amount of the isotope-labeled ABA that is finally detected in each sample. [$^{2}H_{6}$]ABA is commercially available.

References

1. Nambara E, Okamoto M, Tatematsu K, Yano R, Seo M, Kamiya Y (2010) Abscisic acid and the control of seed dormancy and germination. Seed Sci Res 20(2):55–67. https://doi.org/10.1017/S0960258510000012
2. Zhang JH, Jia WS, Yang JC, Ismail AM (2006) Role of ABA in integrating plant responses to drought and salt stresses. Field Crop Res 97(1):111–119. https://doi.org/10.1016/j.fcr.2005.08.018
3. Ton J, Flors V, Mauch-Mani B (2009) The multifaceted role of ABA in disease resistance. Trends Plant Sci 14(6):310–317. https://doi.org/10.1016/j.tplants.2009.03.006
4. Rodriguez PL, Lozano-Juste J, Albert A (2019) PYR/PYL/RCAR ABA receptors. In: Seo M, Marion-Poll A (eds) Abscisic acid in plants, Advances in botanical research, vol 92. Academic Press, London, pp 51–82. https://doi.org/10.1016/bs.abr.2019.05.003
5. Seo M, Marion-Poll A (2019) Abscisic acid metabolism and transport. In: Seo M, Marion-Poll A (eds) Abscisic acid in plants, Advances in botanical research, vol 92. Academic Press, London, pp 1–49. https://doi.org/10.1016/bs.abr.2019.04.004
6. Kuromori T, Seo M, Shinozaki K (2018) ABA transport and plant water stress responses. Trends Plant Sci 23(6):513–522. https://doi.org/10.1016/j.tplants.2018.04.001
7. Krattinger SG, Kang J, Braunlich S, Boni R, Chauhan H, Selter LL, Robinson MD, Schmid MW, Wiederhold E, Hensel G, Kumlehn J, Sucher J, Martinoia E, Keller B (2019) Abscisic acid is a substrate of the ABC transporter encoded by the durable wheat disease resistance gene Lr34. New Phytol 223(2):853–866. https://doi.org/10.1111/nph.15815
8. Pawela A, Banasiak J, Biala W, Martinoia E, Jasinski M (2019) MtABCG20 is an ABA exporter influencing root morphology and seed germination of Medicago truncatula. Plant J 98(3):511–523. https://doi.org/10.1111/tpj.14234
9. Fu X, Liu H, Hassani D, Peng B, Yan X, Wang Y, Wang C, Li L, Liu P, Pan Q, Zhao J, Qian H, Sun X, Tang K (2020) AaABCG40 enhances artemisinin content and modulates drought tolerance in Artemisia annua. Front Plant Sci 11(950):950. https://doi.org/10.3389/fpls.2020.00950
10. Kuromori T, Sugimoto E, Shinozaki K (2021) Brachypodium BdABCG25 is a homolog of Arabidopsis AtABCG25 involved in the transport of abscisic acid. FEBS Lett 595(7):954–959. https://doi.org/10.1002/1873-3468.13925
11. Kanno Y, Hanada A, Chiba Y, Ichikawa T, Nakazawa M, Matsui M, Koshiba T, Kamiya Y, Seo M (2012) Identification of an

abscisic acid transporter by functional screening using the receptor complex as a sensor. Proc Natl Acad Sci U S A 109(24): 9653–9658. https://doi.org/10.1073/pnas.1203567109

12. Shohat H, Illouz-Eliaz N, Kanno Y, Seo M, Weiss D (2020) The tomato DELLA protein PROCERA promotes abscisic acid responses in guard cells by upregulating an abscisic acid transporter. Plant Physiol 184(1):518–528. https://doi.org/10.1104/pp.20.00485
13. Zhang H, Zhu H, Pan Y, Yu Y, Luan S, Li L (2014) A DTX/MATE-type transporter facilitates abscisic acid efflux and modulates ABA sensitivity and drought tolerance in Arabidopsis. Mol Plant 7(10):1522–1532. https://doi.org/10.1093/mp/ssu063
14. Park SY, Fung P, Nishimura N, Jensen DR, Fujii H, Zhao Y, Lumba S, Santiago J, Rodrigues A, Chow TF, Alfred SE, Bonetta D, Finkelstein R, Provart NJ, Desveaux D, Rodriguez PL, McCourt P, Zhu JK, Schroeder JI, Volkman BF, Cutler SR (2009) Abscisic acid inhibits type 2C protein phosphatases via the PYR/PYL family of START proteins. Science 324(5930): 1068–1071. https://doi.org/10.1126/science.1173041
15. Ma Y, Szostkiewicz I, Korte A, Moes D, Yang Y, Christmann A, Grill E (2009) Regulators of PP2C phosphatase activity function as abscisic acid sensors. Science 324(5930): 1064–1068. https://doi.org/10.1126/science.1172408
16. Umezawa T, Sugiyama N, Mizoguchi M, Hayashi S, Myouga F, Yamaguchi-Shinozaki K, Ishihama Y, Hirayama T, Shinozaki K (2009) Type 2C protein phosphatases directly regulate abscisic acid-activated protein kinases in Arabidopsis. Proc Natl Acad Sci U S A 106(41): 17588–17593. https://doi.org/10.1073/pnas.0907095106
17. Boeke JD, LaCroute F, Fink GR (1984) A positive selection for mutants lacking orotidine-5′-phosphate decarboxylase activity in yeast: 5-fluoro-orotic acid resistance. Mol Gen Genet 197(2):345–346. https://doi.org/10.1007/BF00330984
18. Kuromori T, Miyaji T, Yabuuchi H, Shimizu H, Sugimoto E, Kamiya A, Moriyama Y, Shinozaki K (2010) ABC transporter AtABCG25 is involved in abscisic acid transport and responses. Proc Natl Acad Sci U S A 107(5):2361–2366. https://doi.org/10.1073/pnas.0912516107
19. Kang J, Hwang JU, Lee M, Kim YY, Assmann SM, Martinoia E, Lee Y (2010) PDR-type ABC transporter mediates cellular uptake of the phytohormone abscisic acid. Proc Natl Acad Sci U S A 107(5):2355–2360. https://doi.org/10.1073/pnas.0909222107

Chapter 8

Use of Micrografting to Study the Role Played by Peptide Signals in ABA Biosynthesis

Fuminori Takahashi

Abstract

Plants survive by deploying appropriate stress responses under rapidly changing environmental conditions. Vascular plants transmit environmental information among distant organs. Long-distance signaling plays a crucial role in plant adaptation and subsequent survival under severe environmental conditions. In model plants, the micrografting method has recently emerged as an important method to elucidate tissue-to-tissue communication via hormones, RNAs, and peptides. In this chapter, I describe a micrografting method for *Arabidopsis thaliana* to graft shoots onto roots of plants with different genotypes. This grafting method may facilitate further research investigating how vascular plants integrate environmental information among distant organs via long-distance signaling.

Key words Micrografting, Vasculature, Tissue-to-tissue communications, Secretion, Arabidopsis

1 Introduction

Plants have complex and sophisticated molecular mechanisms that rapidly recognize and respond to environmental stress conditions. Vascular plants, in particular, have developed mechanisms for long-distance communications for transmitting environmental information between distant organs [1, 2]. Various methods of plant grafting have been established to analyze the mobile molecules that mediate signal transduction through the vascular bundle under development and stress conditions [3–6]. In Arabidopsis, various types of tissues, including cotyledons, inflorescence stems and young leaves with shoot apical meristem, are subjected to grafting methods [7–11]. Among these tissues, the young shoots and roots of Arabidopsis are used for micrografting within a few days after budding. These grafting experiments are performed using a stereoscopic microscope. The shoot scion and the rootstock heal each other in a few days. The generation of grafted plants over such a short period of time enables us to perform a variety of

Takuya Yoshida (ed.), *Abscisic Acid: Methods and Protocols*,
Methods in Molecular Biology, vol. 2462, https://doi.org/10.1007/978-1-0716-2156-1_8,

analyses at various stages of plant growth. In addition, this useful method can facilitate detailed analyses of various mobile molecules by combining plants with different genotypes.

Recently, micrografting methods have helped researchers elucidate the detailed mechanisms governing tissue-to-tissue communication through various mobile molecules such as phytohormones, RNAs, and peptides under development and stress conditions in Arabidopsis [12–17]. Takahashi et al. indicated that the root-derived CLE25 peptide moves to the leaves to mediate abscisic acid (ABA) biosynthesis and signaling from the roots in response to dehydration stress conditions [18]. This finding helps to elucidate how plants transmit and integrate various biological processes, including development and stress signaling, between distant organs [19, 20]. Therefore, plant grafting methods are important and indispensable techniques for model plant research.

In this chapter, I describe a simple method for Arabidopsis micrografting under nonsterile conditions. This method can generate a grafted plant within a minute, requiring only a degree of manual dexterity and sufficient practice. The grafted plants generated by this method can be utilized for experiments of various purposes on sterile agar medium or soil. This grafting method is an important technique that may help to elucidate new aspects of long-distance signaling in plants.

2 Materials

2.1 Seed Sterilization

1. 70% ethanol.
2. Seed sterilization solution: 1% sodium hypochlorite in sterilized distilled water (sDW).
3. sDW.

2.2 Agar Plate

1. Murashige and Skoog agar medium A: 1/2 Murashige and Skoog (MS), 0.8% agar, without sucrose in polystyrene dishes measuring 140 mm × 100 mm × 14.5 mm in height (*see* **Note 1**).
2. Murashige and Skoog agar medium B: 1/2 MS, 1.5% agar, without sucrose in polystyrene dishes measuring 140 mm × 100 mm × 14.5 mm in height.
3. Murashige and Skoog agar medium C: 1/2 MS, 0.8% agar, 1% sucrose in polystyrene dishes measuring 140 mm × 100 mm × 14.5 mm in height.

2.3 Experimental Supplies for Grafting

1. Filter papers: The filter papers are wrapped in aluminum foil and autoclaved for sterilization (Fig. 1a).

Fig. 1 Preparation of the experimental suppliers for grafting. The filter papers and nitrocellulose membranes are wrapped in aluminum foil and autoclaved for sterilization (**a**). Store the tweezers and culture needle in a 70% ethanol solution (**b**)

2. Nitrocellulose membranes: The membranes are cut into pieces measuring 1 × 3 cm. The membranes are wrapped in aluminum foil and autoclaved for sterilization (Fig. 1a).
3. Tweezers sterilized with 70% ethanol (Fig. 1b, *see* **Note 2**).
4. Stainless culture needles sterilized with 70% ethanol.
5. Tuberculin injection needle (26 G with 1-mL syringe) sterilized by electron beam.
6. Polystyrene dish measuring 90 mm in diameter × 20 mm in height.
7. Stereoscopic microscope.
8. Plant growth chamber.

2.4 Other Supplies and Equipment

1. The seeds of *Arabidopsis thaliana.*
2. Eppendorf tube.
3. Desktop centrifuge.
4. Aluminum foil.
5. Surgical tape.
6. Task wipes.
7. Clean bench.

3 Methods

3.1 Seed Sterilization

1. Add 0.5 mL of 70% ethanol to the Eppendorf tube containing Arabidopsis seeds and mix by inversion. Discard the liquid when the seeds sink to the bottom of the Eppendorf tube.
2. Add 0.5 mL of seed sterilization solution and vortex for 3 min. The Eppendorf tube is gently centrifuged with a desktop centrifuge, and the liquid is subsequently discarded.

Fig. 2 Preparation of Arabidopsis seedlings for grafting. The 1 × 3 cm nitrocellulose membrane is placed on a Murashige and Skoog agar medium A with sterilized tweezers and cultured needles. Place four of the sterilized seeds on each membrane (**a**). The plates are wrapped by task wipe and incubated vertically for 5 days at 22 °C (**b** and **c**)

3. Add 0.5 mL of sDW and vortex for 1 min. The Eppendorf tube is gently centrifuged with a benchtop centrifuge, and the liquid is subsequently discarded. Repeat this washing process 5 times.
4. Add 0.5 mL of sDW. The tube is wrapped in aluminum foil and stored in a refrigerator at 4 °C for at least 2 days.

3.2 Seed Germination

1. The 1 × 3 cm nitrocellulose membrane is placed on a Murashige and Skoog agar medium A with sterilized tweezers and culture needles. The upper and lower membranes should be spaced 2–3 cm apart to ensure that the grown roots do not reach them (Fig. 2a). Approximately 12 membranes can be placed per agar plate on the polystyrene dish.
2. Place four of the sterilized seeds on each membrane.
3. Seal the plate with one to two layers of surgical tape.
4. The plate is allowed to stand and incubate, covered in task wipes, for 5 days at 22 °C (Fig. 2b, c). Covering the plate with task wipes weakens the intensity of light and promotes hypocotyl elongation. The growth chamber space is set at 22 °C with 40 μmol/m^2/s of long-day light (16 h light, 8 h dark).

3.3 Grafting

1. Filter paper is placed on the lid of the polystyrene dish, and sDW is added to the filter paper (Fig. 3a). The filter paper should be kept properly moist to ensure that it does not dry out during the grafting experiment.
2. The nitrocellulose membrane on which the Arabidopsis seedlings were planted is moved from a Murashige and Skoog agar medium A and placed on filter paper. For different genotype grafts, place the membranes in two rows, that is, an upper row and a lower row. The lower row is the membrane on which the plants for shoot genotypes are planted. The upper row contains the plants for root genotype (Fig. 3b).

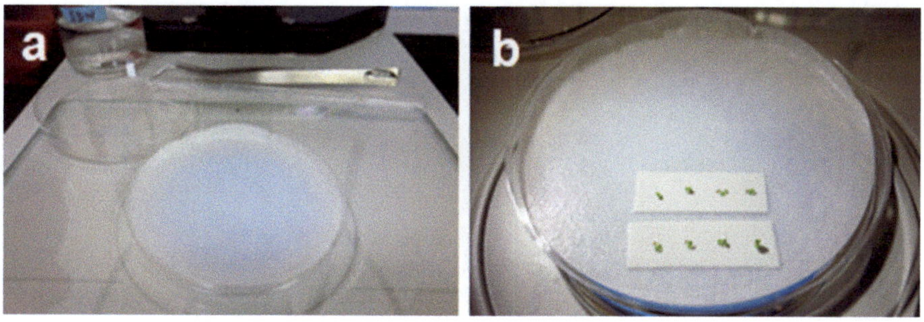

Fig. 3 Setup for the grafting experiments under a stereoscopic microscope. Filter paper is placed on the lid of the polystyrene dish, and sDW is added to the filter paper (**a**). Place the nitrocellulose membrane on which Arabidopsis seedlings are planted on filter paper in two rows, that is, an upper row and a lower row (**b**)

Fig. 4 Cutting of Arabidopsis to create rootstocks. The top row of the membrane contains the plants for root genotype (**a**). The hypocotyl is cut into round slices, and the shoot is removed using a tuberculin injection needle (**b**). Cut the hypocotyl vertically to make it Y-shaped (**c**). The dashed lines indicate where the Arabidopsis hypocotyl is cut to create rootstock

3. Several drops of sDW are added near the membranes to maintain a constant level of membrane moistness during the grafting experiments (*see* **Note 3**).
4. First, rootstocks are created with the top row of the membrane under a stereoscopic microscope (*see* **Note 4**). The hypocotyl is cut into round slices, and the shoot is removed using a tuberculin injection needle (Fig. 4a, b, *see* **Note 5**). The hypocotyl is cut vertically to make it Y-shaped (Fig. 4c, *see* **Note 6**).
5. Next, the shoot scions are created with the bottom row of the membrane. Cut the hypocotyl diagonally (Fig. 5a, b, *see* **Note 7**).
6. Carry the shoot scion to the side of the rootstock using the sterilized culture needle (Fig. 6a).
7. Connect the hypocotyl cross section of the shoot scion into a V-shaped notch of the rootstock (Fig. 6b). Check the hypocotyl junction at different angles to ensure that there is no misalignment. Perform the **steps 4–7** for one grafted plant within 1 min (*see* **Note 8**).

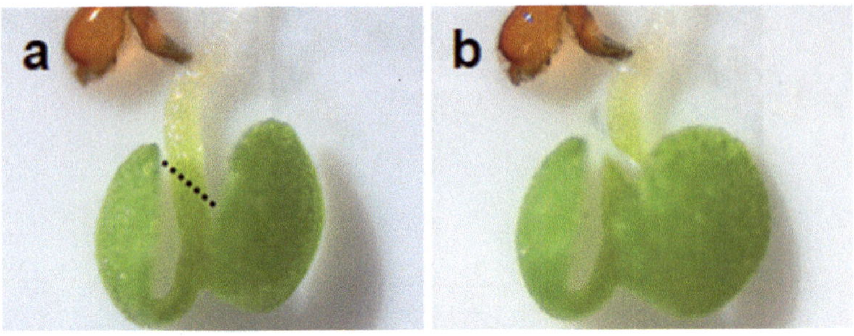

Fig. 5 Cutting of Arabidopsis to create shoot scions. The bottom row of the membrane contains the plants for shoot genotype (**a**). Cut the hypocotyl diagonally (**b**). The dashed line indicates where the Arabidopsis hypocotyl is cut to create shoot scion

Fig. 6 Connect the shoot scion onto the rootstock. Carry the shoot scion to the side of the rootstock using the sterilized culture needle (**a**). Join the hypocotyl cross section of the shoot scion into a V-shaped notch of the rootstock (**b**). Place the Murashige and Skoog agar medium B at an angle in the plant growth chamber and incubate at 27 °C for 4 days (**c**)

8. Transfer the membrane that contains the grafted plants to the Murashige and Skoog agar medium B immediately after grafting all the plants (e.g., four grafted plants) on the membrane (*see* **Note 9**). Close the lid of the plate and keep it still. Approximately 6 membranes can be placed per agar plate.
9. The plate is sealed with one to two layers of surgical tape, placed at an angle (approximately 30°) in the plant growth chamber, and incubated at 27 °C to promote the concrescence of the hypocotyl junction for 4 days (Fig. 6c).
10. Transfer the plate to another growth chamber and incubate it at 22 °C for 2–4 days (*see* **Note 10**).
11. Adventitious root formation is checked with a stereoscopic microscope (Fig. 7a). The adventitious roots generated from the junction of grafting are cut using a tuberculin injection needle (Fig. 7b).

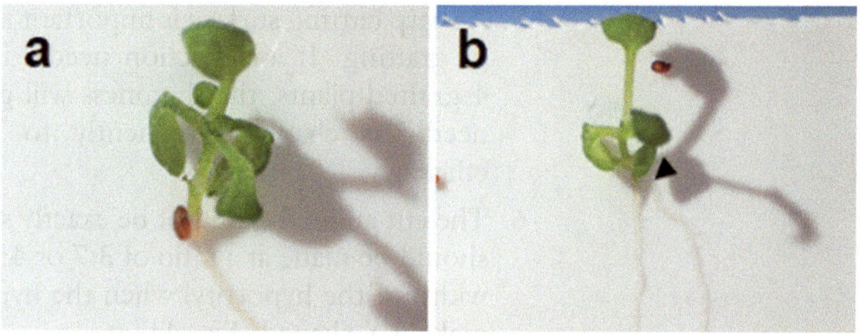

Fig. 7 Check adventitious root formation with a stereoscopic microscope. The grafted plant that does not have any adventitious roots (**a**). The black arrow indicates the adventitious roots that are generated from the junction of grafting (**b**)

12. Transplant the grafted plants onto a Murashige and Skoog agar medium C on a clean bench and subsequently incubate them at 22 °C for 2–4 days (*see* **Note 11**).
13. Transfer the grafted plants to media or soil for further experiments.

4 Notes

1. It suppresses the growth of miscellaneous bacteria, even when grafting is performed under nonsterile conditions, since it does not contain sucrose.
2. Store the tweezers and culture needles in a 70% ethanol solution. In addition, the tweezers and culture needles are dipped in 70% ethanol solution periodically during the grafting experiment to reduce contamination under nonsterile conditions.
3. Membrane moisture is an important factor for successful grafting. The moisture should be kept constant throughout the grafting experiments. It is preferred that the roots and seed coats are covered with a film of water, while the hypocotyls are not submerged in water. When the hypocotyl is submerged in water, the vascular fluid tends to flow out, causing a decrease in the grafting success rate. This phenomenon will also reduce the grafting success rate if the membrane is too dry. The humidity of the room is 50% or more, and the temperature is approximately 22 °C, which also contributes to the efficiency of grafting experiments.
4. The direction and angle of the tip of the injection needle should be considered when cutting. Hold the culture needle in the other hand and use it to support the plants and injection needles.

5. A sharp cutting surface is important to increase the success rate of grafting. If an injection needle is used on approximately 4 grafted plants, the sharpness will gradually deteriorate. The needle is changed frequently to avoid reducing grafting efficiency.
6. The cut angle should not be exactly straight. Ideally, the notch should be made at a ratio of 3:7 or 4:6 from left to right to the width of the hypocotyl when the hypocotyl is cut vertically to make it Y-shaped (Fig. 4a, c).
7. Note that the directions of the notch (**step 4**) and the diagonal cut (**step 5**) should be the same. If the work is performed in the opposite direction, then the notch ratio of the rootstock and the diagonal cutting direction of the shoot scion are also all opposite.
8. If the grafting experiment is successful, the cut surfaces will stick together due to the sharply cut surface and the viscosity of the vascular fluid. The graft will not come off easily, even if the plant is poked with a needle.
9. The membrane is carefully transferred from the filter paper to the agar plate to ensure that the grafted joints do not shift.
10. Continue to place the plate at an angle (approximately 30 degrees) in the plant growth chamber.
11. Incubate the plate vertically in the plant growth chamber.

Acknowledgments

This work was supported by JSPS KAKENHI Grant Number JP19H03255 (to F.T.), JP20K21437 (to F.T.) and by research grants from the Toray Science Foundation (to F.T.). I appreciate the valuable comments of Dr. Kazuo Shinozaki of the RIKEN Center for Sustainable Resource Science.

References

1. Takahashi F, Kuromori T, Urano K, Yamaguchi-Shinozaki K, Shinozaki K (2020) Drought stress responses and resistance in plants: from cellular responses to long-distance intercellular communication. Front Plant Sci 11:556972. https://doi.org/10.3389/fpls.2020.556972
2. Yoshida T, Fernie AR, Shinozaki K, Takahashi F (2021) Long-distance stress and developmental signals associated with abscisic acid signaling in environmental responses. Plant J 105(2):477–488. https://doi.org/10.1111/tpj.15101
3. Melnyk CW, Meyerowitz EM (2015) Plant grafting. Curr Biol 25(5):R183–R188. https://doi.org/10.1016/j.cub.2015.01.029
4. Melnyk CW, Schuster C, Leyser O, Meyerowitz EM (2015) A developmental framework for graft formation and vascular reconnection in Arabidopsis thaliana. Curr Biol 25(10): 1306–1318. https://doi.org/10.1016/j.cub.2015.03.032
5. Cookson SJ, Ollat N (2013) Grafting with rootstocks induces extensive transcriptional re-programming in the shoot apical meristem

of grapevine. BMC Plant Biol 13:147. https://doi.org/10.1186/1471-2229-13-147

6. Turnbull CG, Booker JP, Leyser HM (2002) Micrografting techniques for testing long-distance signalling in Arabidopsis. Plant J 32(2):255–262. https://doi.org/10.1046/j.1365-313x.2002.01419.x
7. Nisar N, Verma S, Pogson BJ, Cazzonelli CI (2012) Inflorescence stem grafting made easy in Arabidopsis. Plant Methods 8(1):50. https://doi.org/10.1186/1746-4811-8-50
8. Huang NC, Yu TS (2015) A pin-fasten grafting method provides a non-sterile and highly efficient method for grafting Arabidopsis at diverse developmental stages. Plant Methods 11:38. https://doi.org/10.1186/s13007-015-0081-7
9. Melnyk CW (2017) Grafting with Arabidopsis thaliana. Methods Mol Biol 1497:9–18. https://doi.org/10.1007/978-1-4939-6469-7_2
10. Huang NC, Yu TS (2021) Non-sterile grafting methods for Arabidopsis. Methods Mol Biol 2200:113–119. https://doi.org/10.1007/978-1-0716-0880-7_4
11. Yoo SJ, Hong SM, Jung HS, Ahn JH (2013) The cotyledons produce sufficient FT protein to induce flowering: evidence from cotyledon micrografting in Arabidopsis. Plant Cell Physiol 54(1):119–128. https://doi.org/10.1093/pcp/pcs158
12. Tabata R, Sumida K, Yoshii T, Ohyama K, Shinohara H, Matsubayashi Y (2014) Perception of root-derived peptides by shoot LRR-RKs mediates systemic N-demand signaling. Science 346(6207):343–346. https://doi.org/10.1126/science.1257800
13. Ohkubo Y, Tanaka M, Tabata R, Ogawa-Ohnishi M, Matsubayashi Y (2017) Shoot-to-root mobile polypeptides involved in systemic regulation of nitrogen acquisition. Nat Plants 3:17029. https://doi.org/10.1038/nplants.2017.29
14. Ota R, Ohkubo Y, Yamashita Y, Ogawa-Ohnishi M, Matsubayashi Y (2020) Shoot-to-root mobile CEPD-like 2 integrates shoot nitrogen status to systemically regulate nitrate uptake in Arabidopsis. Nat Commun 11(1):641. https://doi.org/10.1038/s41467-020-14440-8
15. Chen X, Yao Q, Gao X, Jiang C, Harberd NP, Fu X (2016) Shoot-to-root mobile transcription factor HY5 coordinates plant carbon and nitrogen acquisition. Curr Biol 26(5):640–646. https://doi.org/10.1016/j.cub.2015.12.066
16. Osugi A, Kojima M, Takebayashi Y, Ueda N, Kiba T, Sakakibara H (2017) Systemic transport of trans-zeatin and its precursor have differing roles in Arabidopsis shoots. Nat Plants 3:17112. https://doi.org/10.1038/nplants.2017.112
17. Notaguchi M, Kurotani KI, Sato Y, Tabata R, Kawakatsu Y, Okayasu K, Sawai Y, Okada R, Asahina M, Ichihashi Y, Shirasu K, Suzuki T, Niwa M, Higashiyama T (2020) Cell-cell adhesion in plant grafting is facilitated by beta-1,4-glucanases. Science 369(6504):698–702. https://doi.org/10.1126/science.abc3710
18. Takahashi F, Suzuki T, Osakabe Y, Betsuyaku S, Kondo Y, Dohmae N, Fukuda H, Yamaguchi-Shinozaki K, Shinozaki K (2018) A small peptide modulates stomatal control via abscisic acid in long-distance signalling. Nature 556(7700):235–238. https://doi.org/10.1038/s41586-018-0009-2
19. Takahashi F, Shinozaki K (2019) Long-distance signaling in plant stress response. Curr Opin Plant Biol 47:106–111. https://doi.org/10.1016/j.pbi.2018.10.006
20. Takahashi F, Hanada K, Kondo T, Shinozaki K (2019) Hormone-like peptides and small coding genes in plant stress signaling and development. Curr Opin Plant Biol 51:88–95. https://doi.org/10.1016/j.pbi.2019.05.011

Chapter 9

Gas Exchange Measurements in Systemic Signaling Studies

Sanna Ehonen and Maija Sierla

Abstract

Different parts of a plant can be simultaneously exposed to very different conditions, for example a leaf moving in and out of shadow. In addition to local responses, transmission of information between different tissues and organs is thought to affect the coordination of overall responses to changing environmental conditions. An important adaptive role is played by the stomata, which regulate the evaporation of water vapor and supply of CO_2 for photosynthesis. Here, we describe a method to study the effect of distally triggered systemic signals on stomatal conductance. The experimental set up, consisting of a growth chamber and a leaf gas exchange measuring system, enables time-resolved measurements on an intact leaf while maintaining a full control over the environmental conditions of the measured leaf and the whole seedling. The method can be used as a powerful tool to study short- and long-term stomatal responses to changes in different environmental variables, such as light.

Key words Stomata, Systemic signaling, Stomatal conductance, Photosynthesis, Leaf gas exchange

1 Introduction

Stomata sense and respond to changes in different environmental factors, including light, ambient CO_2 concentration, air humidity, and temperature. High CO_2 and dark treatments are commonly used to elicit rapid stomatal closure and there are several *Arabidopsis thaliana* mutants with impaired responses to these treatments. Mutant plants lacking the function of the key components of guard cell abscisic acid (ABA) [1] and CO_2 signaling [2, 3] pathways are impaired in rapid stomatal closure triggered by darkness or elevated CO_2. Although several different studies (mainly on Arabidopsis) have identified key molecular components involved in guard cell signaling pathways, a more holistic understanding of the mechanisms that underlie stomatal function at the whole plant level is largely missing.

Plants respond to abiotic stress through a systemic response termed systemic acquired acclimation (SAA) [4, 5]. Characteristic to this response is a rapid spread of signals to distal parts of the plant

Takuya Yoshida (ed.), *Abscisic Acid: Methods and Protocols*,
Methods in Molecular Biology, vol. 2462, https://doi.org/10.1007/978-1-0716-2156-1_9,

within minutes after the application of a treatment to a local leaf. Signals are transmitted through several different mechanisms including reactive oxygen species (ROS) waves [6], calcium (Ca^{2+}) waves [7], and electrical signals [8]. Heat stress, wounding, high light, darkness, ABA and increased CO_2 treatments on local leaves were shown to trigger rapid systemic signaling that is able to elicit stomatal closure in nontreated distal leaves in several different species [9–17]. However, these studies were conducted with different techniques, treatments and plant species and the evidence for the role of rapid systemic signals in stomatal regulation is mixed, especially for Arabidopsis. Recently, stomatal aperture measurements performed on epidermal peels of low light-grown Arabidopsis plants indicated that systemic signals regulate stomatal aperture in distal leaves in response to high light, wounding and ABA [14, 15]. Conversely, in our recent study using stomatal conductance measurements of intact leaves, systemic signals triggered a decrease in stomatal conductance in response to darkness and high CO_2 treatments in two different tree species, ***Populus tremula*** × ***tremuloides*** and ***Betula pendula***, but not in Arabidopsis [16]. Also, in a recent study using thermal imaging as a tool to study systemic stomatal responses, local application of ABA did not trigger fast stomatal closure in distal leaves of Arabidopsis [18].

Leaf gas exchange, whole plant gas exchange, thermal imaging, and epidermal peels are widely used methods for studying stomatal responses to changes in different environmental variables. As each method has its limitations, these should be carefully considered when designing an experiment and selecting the method. Gas exchange measurements have several advantages in systemic signaling studies: (a) the whole leaf can be measured at once, (b) the ambient conditions of the measured leaf can be tightly controlled (c) noninvasive and high precision detection of changes in stomatal function, and (d) ability to perform time-resolved measurements on the same leaf.

Here we describe a simple and efficient method to study the involvement of systemic signals on photosynthesis and stomatal conductance in different species. In the unique experimental set up two gas exchange measuring devices are placed inside a growth chamber allowing tight control over the environmental conditions of the whole plant. Treatments can be applied on a single leaf or the whole plant in different combinations in a noninvasive and controlled manner while minimizing the occurrence of possible artifacts. The method can be used to study the effect of different environmental variables on short- and long-term responses in leaf gas exchange.

2 Materials

2.1 Equipment

1. Leaf gas exchange measuring system with effective and stable control of temperature, air humidity, CO_2 and light, for example Heinz Walz GFS3000 or equivalent (*see* **Note 1**).
2. Controlled environment chambers with effective and stable additive and subtractive control of air humidity and CO_2 concentration, for example Aralab FitoClima Growth Chamber or equivalent (*see* **Note 2**).

2.2 Plant Material

1. Outdoor grown hybrid aspen (*Populus tremula* × *tremuloides*) seedlings treated with an insecticide: Use different genotype for each biological replicate.
2. Outdoor grown birch (*Betula pendula*) seedlings treated with an insecticide: Use different genotype for each biological replicate.
3. Arabidopsis (*Arabidopsis thaliana*) plants of Col-0 accession grown in 1:1 peat–vermiculite mixture: Use seedlings of the same growth stage.

2.3 Computational Environment

1. The mgcv package (version 1.8-17) in R (version 3.4.2; R core team2017; http://www.r-project.org/) for statistical analysis. Other statistical software or methods may also be used.

3 Methods

3.1 Preparing Tree Seedlings

1. Use outside grown, hybrid aspen and birch seedlings.
2. Wash the aerial parts of the seedlings with water and spray with insecticide immediately prior to acclimatization period.
3. Acclimatize the seedlings in the growth chamber for 1 week under conditions similar to their natural environment. We use: 20 h/4 h (day/night) photoperiod, average PAR 600 μmol/m^2/s, relative humidity 60%/70% (day/night), temperature 20 °C/16 °C (day/night), and CO_2 concentration of 400 ppm. These are also the control conditions during the treatments.
4. Select a mature leaf positioned just below the top third of the canopy and place it in the cuvette of the gas exchange measuring device. Use leaves with similar leaf morphology. Make sure that also the leaf margins are inside the cuvette.

3.2 Preparing Arabidopsis Seedlings

1. Sow Arabidopsis seeds in 1:1 peat–vermiculite mixture and keep in 4 °C in the dark for 2–3 days.
2. Grow plants for at least 4 weeks in the growth chamber under the following conditions: 12 h/12 h (day/night) photoperiod,

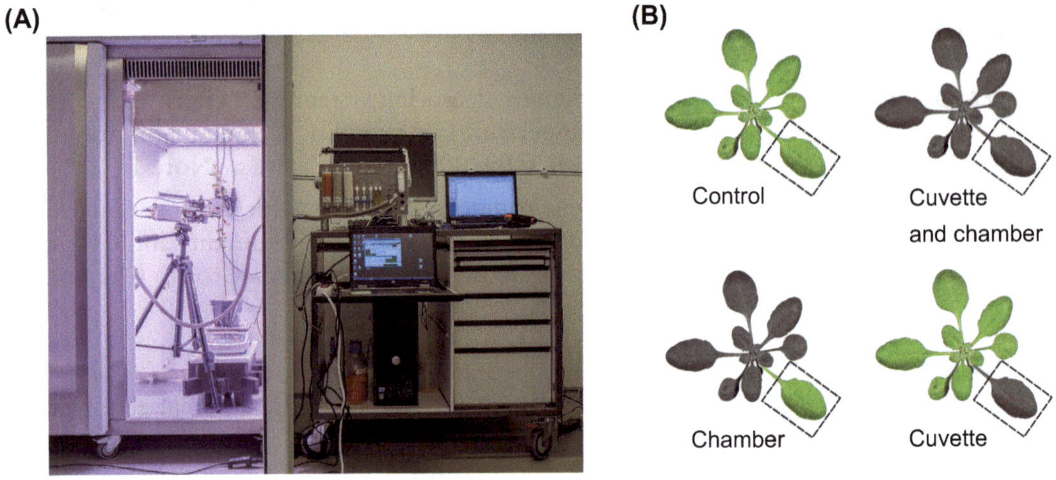

Fig. 1 Experimental system and treatment combinations. (**a**) Experimental setup where a seedling and the measuring head of a gas exchange measuring device (continuously measuring one leaf) are placed inside a growth chamber. (**b**) Schematic representation of treatment combinations: control, treatment in cuvette and chamber, treatment only in chamber and treatment only in cuvette. Black color indicates the treated area and boxes with dashed borders illustrate the cuvette of a gas exchange measuring device

average PAR 220 μmol/m^2/s, relative humidity 60%/70% (day/night) temperature 22 °C/18 °C (day/night), and CO_2 concentration of 400 ppm. These are also the control conditions during the treatments.

3. Count the leaves according to the Arabidopsis phyllotaxy (*see* **Note 3**). Select a fully expanded leaf that can be fitted to the cuvette and use the same leaf from each measured rosette.

3.3 Measuring Setup (Fig. 1a)

1. Use two identical growth chambers for the experiment; one for growing/acclimatizing the plants (growth chamber) and the other for performing the treatments (experimental chamber).
2. Place the two identical gas exchange measuring devices with measuring heads (cuvettes) inside the experimental chamber allowing measurement of two plants simultaneously (*see* **Note 4**).
3. Place the central units of the gas exchange measuring devices outside the growth chambers (*see* **Note 5**).
4. Place one leaf of the experimental plant inside a cuvette of the gas exchange measuring device. Make sure that the whole leaf including the leaf margins is enclosed within the cuvette.

3.4 Treatments and Measurements

1. Carry out all measurements between 10.00 am and 16.00 pm in random order (*see* **Note 6**).
2. Transfer two plants from the growth chamber into the treatment chamber (*see* **Notes 7** and **8**).

3. Place one leaf of each plant carefully in the measuring cuvettes. Minimize handling of the plants (*see* **Note 7**) and breathing inside the chamber (*see* **Note 8**).
4. Allow a minimum of 1 h acclimation period in the species specific growth conditions (Subheadings 3.1 and 3.2; *see* **Note 9**).
5. Treat and measure two plants simultaneously with two different gas exchange measuring devices (*see* **Note 1**).
6. Apply the darkness or rapid increase in CO_2 concentration–treatment (*see* **Notes 10** and **11**) in one of the four different combinations; (1) Apply the treatment to the measured leaf only (cuvette), (2) Apply the treatment to the rest of the seedling (chamber) excluding the leaf within the cuvette, (3) Apply the treatment to the entire seedling (cuvette and chamber), (4) Do not apply treatment (control). Seedling must be changed after every treatment or control measurement. Apply the treatments in a different and random order to each plant and allow sufficient recovery between treatments (*see* **Notes 10–12**).
7. Record the gas exchange parameters at least 15 min prior to the onset of the treatment and at least 45 min after the onset of the treatment, depending on the research question (*see* **Note 13**).
8. Perform at least four biological replicates of each treatment (*see* **Note 12**).
9. Make recordings of gas exchange data at short time intervals, for example one measurement every 5 s, and perform zero point calibrations as advised by the manufacturer (*see* **Notes 14** and **15**).

3.5 Statistical Analysis and Evaluation of the Results

1. Normalize the relative stomatal conductance (g_s), CO_2 assimilation (A) and intercellular CO_2 concentration (c_i) values to the pretreatment steady state value (time point 0).
2. Perform data exploration with statistical techniques and data visualization (*see* **Note 16**).
3. Analyze the leaf responses to the applied treatments by fitting generalized additive mixed models (GAMMs) to g_s, A and c_i as functions of time using mgcv package in R (*see* **Note 17**) in **steps 4–8**.
4. Estimate trends in the selected response (g_s, A or c_i) to each treatment combination (Subheading 3.4) for selected species by fitting GAMM with gam function of the mgcv package, using the following model:

```
out.gamm <- gam (Y ~ Treatment + s(Time) + s(Time, by=Treat-
mentOF) + s(Subject, Time, bs="re", by=dummy), method="REML")
```

This model estimates the average g_s, A or c_i (Y) between the different treatments (Treatment) over a time course. (Treatment + s(Time) + s(Time, by=TreatmentOF)) calls for different trends over time for different treatment combinations by using an ordered factor. (s(Subject, Time, bs="re")) calls for random slopes for the trend of each subject (*see* **Note 18**).

5. Evaluate the model by checking the distribution and autocorrelation of the residuals.
6. For fitting the model, use dummy variable set to 1.
7. For data visualization, set the dummy variable to zero to get predictions with the prediction subject effects set to zero.
8. Do multiple comparisons of the treatment responses by reordering the levels of the ordered factor (TreatmentOF) (*see* **Note 19**).
9. Compare the differences in smooth functions for selected pairs of treatments using a prediction matrix (*see* **Note 20**) in **steps 10–12**.
10. Fit GAMM to the data (using the model without an ordered factor):

```
out.gamm <- gam (Y ~ Treatment + s(Time, by=Treatment) + s
(Subject, Time, bs="re", by=dummy), method="REML")
```

11. Evaluate the model by checking the distribution and autocorrelation of the residuals.
12. Calculate the differences between pairs of the four smooths estimated in the model (i.e., the four treatment combinations, Subheading 3.4) via the use of a prediction matrix.
13. Compare and evaluate the models with the original data (*see* **Notes 21** and **22**).

4 Notes

1. Precise measurements of stomatal conductance and photosynthesis of the distal leaf require a gas exchange measuring device with a closed chamber or cuvette where a leaf (or a branch) can be placed without detaching it from the rest of the plant. The system must be able to control the environmental quantities important to photosynthesis: CO_2, light, humidity, and temperature. We use portable gas exchange measuring systems (model no. GFS3000, Heinz Walz) which enable precise control of the important environmental variables in the measuring cuvette and automatic execution of the measuring program. The experiment can be done with a single gas exchange

measuring device, however, two (or more) devices facilitate a more reasonable time frame for the experiments. The handling and measurement of one seedling takes a minimum of 2 h and in our experience the maximum number of successful measurements per day with one gas exchange device is two.

2. In order to control the ambient environment of the whole plant, growth chambers are required. We use two identical chambers; one for growing the plants and the other for performing experiments. The chambers must be able to control the same environmental variables as the gas exchange measuring system. Minimum requirement is the ability to control the studied environmental variables, as the whole plant treatments (in growth chamber) must be performed under the same conditions as the local treatments (in gas exchange measuring device). If the chamber does not have CO_2 control, an external CO_2 measuring instrument should be used to monitor the CO_2 concentration inside the chamber (*see* **Note 8**). We use customized FitoClima 1200 Growth Chambers (Aralab) with the ability to control CO_2, temperature, humidity and light. In addition, the chambers have B50 led lights (e.g., Valoya Oy or equivalent) which enable the use of relatively high light intensities.
3. Start counting from the oldest leaf, that is, leaf 1 being the oldest leaf in the rosette [19].
4. The devices can also be used to measure and control the ambient environment of two separate leaves of the same plant.
5. This allows the user to operate the system without opening the growth chambers during the measurements.
6. Stomatal conductance and responsiveness is dependent on the time of the day. When working with Arabidopsis start the measurements at least 2 h after the onset of the light period and perform them for a maximum of 6 h/day. When working with outside grown tree seedlings, attention must be paid to the circadian and diurnal rhythms of photosynthesis and stomatal conductance. Use the same day length that the seedlings experienced when growing outside. We recommend to test the daily rhythms of leaf gas exchange and stomatal responsiveness of the selected species before the onset of the experiment.
7. Handling of a plant is known to trigger stomatal responses which in Arabidopsis have been shown to require a recovery period of 30–60 min before cellular/molecular processes can be fully reactivated [20]. Extra care must be taken when tightening the cuvette; too much force/pressure may produce lesions on leaf tissue and inhibit the transduction of systemic signals. It is generally good practice to inspect the measured

leaf 24 h after measurements, especially when working with species with tender leaves, such as Arabidopsis. If lesions or wounds occur, the measurements must be discarded.

8. The time spent inside the chamber when replacing the experimental plants should be kept to minimum as the presence of the researcher affects the conditions in the chamber. Importantly, breathing inside the chamber should be avoided if possible as this increases the CO_2 concentration rapidly and for an extended period despite the capability of the chamber to control CO_2 levels. This can have a significant impact on the outcome of the study.

   To exemplify and highlight this point, we conducted a simple experiment (Fig. 2). Ambient CO_2 concentration of a 15 m^2 air-conditioned room and stomatal conductance and assimilation of an Arabidopsis plant inside the room were measured during a visit of a researcher. Ambient CO_2 concentration increase in the room and the associated effects on stomatal conductance and assimilation of the plant were evident within minutes of the entry of the researcher (Fig. 2).

   As the ambient CO_2 concentration affects both local and systemic stomatal responses, it is essential to monitor these values when conducting systemic signaling studies. Otherwise it is not possible to exclude the possibility that the observed responses are triggered by the presence of the researcher instead of the applied stimuli. Such artifacts are minimized in the experimental setup described here as the automatized treatments and measurements inside a chamber do not require the presence of the researcher.

9. The acclimation period should be started once the chamber conditions have reached the control conditions after exit of the researcher.

10. Ambient conditions can usually be controlled very precisely in a cuvette of a gas exchange measuring device but not as efficiently in growth chambers due to the larger volume of the chamber. Therefore, the studied environmental variables largely depend on the speed and precision with which the growth chamber can control and adjust the climatic conditions. Both the growth chamber and gas exchange measuring system allow automatic control of CO_2, light, temperature and humidity. However, only the increase in CO_2 concentration and changes in light level can be used as treatments under our current set up. Changes in humidity or decrease in ambient CO_2 concentration in the growth chamber cannot be regulated in the required speed, accuracy and reproducibility.

11. We used sudden darkness during the light period and rapid increase in ambient CO_2 concentration as treatments. The increase in CO_2 concentration from 400 to 1100 ppm was

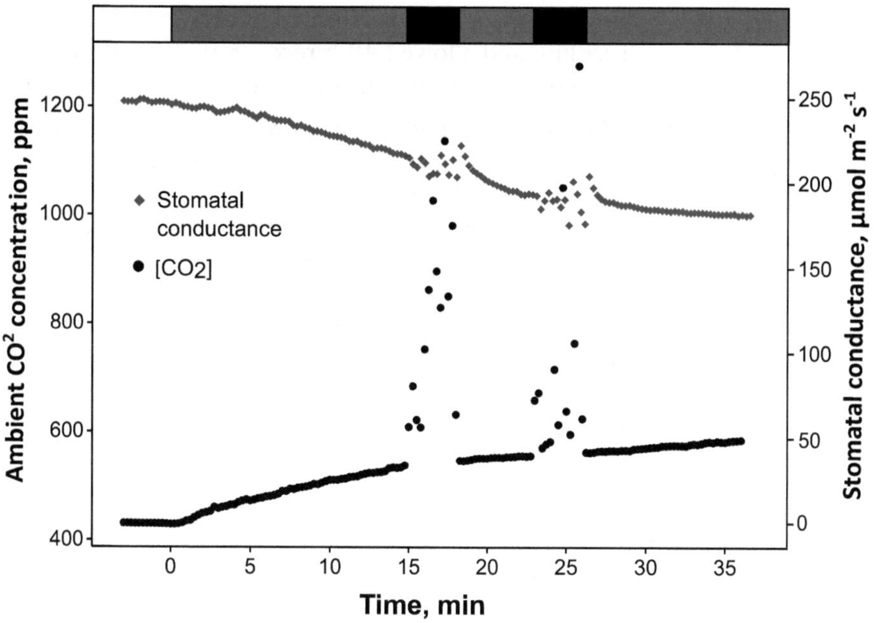

Fig. 2 Representative figure on the effect of increased room CO_2 levels (ppm, black circles) on the stomatal conductance (μmol/m^2/s, gray diamonds) of Arabidopsis. An experiment was performed to test whether the presence of a researcher affects the room CO_2 concentration and consequently whole plant gas exchange. A 1-h acclimation period in ambient conditions (PAR 300 μmol/m^2/s, relative humidity 40%, temperature 22 °C, and CO_2 concentration 430 ppm) was followed by one person entering the room (time point 0). The stomatal conductance of Arabidopsis rosette was continuously measured with a gas exchange measuring device, GFS3000 (Heinz Walz GmbH, Germany), which was set to follow the ambient temperature, humidity and CO_2 concentration. The person in the room kept at least 1.5 m distance to the air inlet of the gas exchange measuring system (gray bar) except for two 3-min time periods when the person stayed at approximately 30 cm distance to the air inlet (black bar). The periods at close distance resemble a situation when a researcher would be applying a treatment or sampling the plant. The results indicate that the increase in ambient CO_2 concentration caused by the presence of the researcher has a significant effect on stomatal conductance of the measured plant. Interestingly, the presence of the researcher at close proximity to the air inlet had a balancing effect on stomatal conductance. This was likely due to transient decrease in vapor pressure deficit during the close proximity treatment (data not shown). The experiment exemplifies that the presence of a researcher in the room where an experiment is performed may have unwanted effects on stomatal conductance and result in false positive results

reached on average in 4.5 min. If the system used is capable of a faster and slightly smaller increase in CO_2 concentration, this is advisable.

12. Note that if the treatment is applied to the whole growth chamber, the two plants measured with different gas exchange measuring devices are not replicates of the same species, genotype or ecotype. Use different tree genotype for each biological replicate. The same tree seedling can be used for the different treatment conditions and combinations as long as sufficient recovery time is allowed between experiments. Due to limited

growth space, we performed seven experiments with each seedling and allowed 48 h recovery between experiments. For Arabidopsis, a different plant was used for every experiment.

13. The time required for measurements depends on the species used, applied treatments, studied responses (g_s, A or both), and the signaling pathways of interest. Rapid systemic signals transmitted by ROS, Ca^{2+}, and electric waves can be expected to trigger fast responses (in the order of a few minutes) whereas slower signals, such as changes in hydraulic and osmotic potentials and phytohormone concentration, may result in responses observed several minutes or hours after the onset of the treatment [16]. It is generally good practice to conduct preliminary experiments, where stomatal conductance and assimilation are measured for a few hour after the onset of the applied treatments.

14. Even a slight difference in temperature between the chamber and the cuvette can cause condensation within the gas exchange measuring system resulting in increased noise in the conductance measurements; rapid fluctuations of above 10 $mmol/m^2/s$ are typically indicative of this. This problem can be avoided by drying the system regularly, which in our setup was accomplished by circulating dry air through the system overnight every second night.

15. In order to gain detailed information on the kinetics of the gas exchange responses and to detect possible malfunctions in the measuring system, we recommend to do the measurements at short time intervals.

16. Data exploration is an essential, initial step in data analysis. It helps to identify the nature of the data, the relationships between variables and the presence of outliers. Violations of the underlying assumptions of the applied statistical analysis may result in considerable errors in statistical hypothesis testing and lead to false conclusions. Detailed data exploration helps to identify potential problems and can be conducted by using several different statistical techniques and data visualization [21].

17. Time course data, where individual data points are from the same sample cannot be analyzed by simple statistical methods such as ANOVA as the data violates the basic assumption of independent observations. Generalized additive mixed modelling (GAMM) [22] is an appropriate and powerful statistical method to analyze time series data and to test statistical differences between the treatment responses. The method does not assume the data to be linear but instead determines the degree of linearity or nonlinearity from the data. In addition, GAMM allows the inclusion of random effects (e.g., subject) and thus

takes into account the nonindependence of the measuring points. In gas exchange data the individual measuring points are from the same subject (plant) and the data has a nested structure. In ecological research such methods are already widely used but to our knowledge only a few studies have thus far used GAMM or similar statistical methods for the analysis of gas exchange data.

18. The model incorporates a separate trend for each of the four treatment combination (Subheading 3.4) into a single model of temporal changes in the selected gas exchange parameter. For more detail *see* [22]. Fit a model separately for (1) both treatment conditions (i.e., darkness and CO_2, Subheading 3.4), (2) each species (i.e., hybrid aspen, birch and Arabidopsis; Subheading 2.2), and (3) each gas exchange parameter (i.e., g_s, A and c_i; Subheading 3.5)
19. Multiple comparisons require the adjustment of *p*-value by Bonferroni correction (with 12 different treatment combinations the adjusted *p*-value is 0.004). Significant *p*-value (<0.004) indicates that the two responses followed a different trend over time. Setting dummy variable to zero allows the predictions with the subject effect set to zero (i.e., to fit GAMM model for the data of each treatment group). The use of different factor levels as reference level may result in two different *p*-values for the same comparisons. In these cases use the *p*-value of the best model fit (ranked according to their AIC values).
20. This is a more specific method for comparing the differences between the treatment responses. It allows identification of the specific time period at which the smooths were significantly different from each other. For more detailed description *see* [22, 23].
21. GAMM models are often presented without original data as the model is considered a sufficient representation of the data. Likewise, gas exchange data is mostly presented as mean value with standard error. Publishing the original data together with the model/mean values is advisable as it shows the true variation of the data that can be meaningful for interpretation of the results.
22. From designing the experiment to data analysis and interpretation it should be kept in mind that statistical significance does not always imply biological relevance. In the setup described here the application of two different treatments (CO_2 and darkness) in three different combinations on different species increases the biological relevance of the results.

Acknowledgments

We thank Mikael Brosché and Pedro Aphalo for fruitful discussions and critical comments on the manuscript and Tuomas Puukko for providing the pictures. The authors were supported by the Academy of Finland (grants 333703 and 336359 to M.S.), Ella and Georg Ehrnrooth Foundation (M.S.), Finnish Cultural Foundation (M.S.), and Alfred Kordelin Foundation (S.E.).

References

1. Merilo E, Laanemets K, Hu H et al (2013) PYR/RCAR receptors contribute to ozone-, reduced air humidity-, darkness-, and CO2-induced stomatal regulation. Plant Physiol 162:1652–1668. https://doi.org/10.1104/pp.113.220608
2. Hashimoto M, Negi J, Young J et al (2006) Arabidopsis HT1 kinase controls stomatal movements in response to CO2. Nat Cell Biol 8:391–397. https://doi.org/10.1038/ncb1387
3. Hõrak H, Sierla M, Tõldsepp K et al (2016) A dominant mutation in the HT1 kinase uncovers roles of MAP kinases and GHR1 in CO2-induced stomatal closure. Plant Cell 28:2493–2509. https://doi.org/10.1105/tpc.16.00131
4. Karpinski S, Reynolds H, Karpinska B et al (1999) Systemic signaling and acclimation in response to excess excitation energy in Arabidopsis. Science 284:654–657. https://doi.org/10.1126/science.284.5414.654
5. Suzuki N, Miller G, Salazar C et al (2013) Temporal-spatial interaction between reactive oxygen species and abscisic acid regulates rapid systemic acclimation in plants. Plant Cell 25:3553–3569. https://doi.org/10.1105/tpc.113.114595
6. Miller G, Schlauch K, Tam R et al (2009) The plant NADPH oxidase RBOHD mediates rapid systemic signaling in response to diverse stimuli. Sci Signal 2:ra45. https://doi.org/10.1126/scisignal.2000448
7. Choi W-G, Toyota M, Kim S-H et al (2014) Salt stress-induced Ca2+ waves are associated with rapid, long-distance root-to-shoot signaling in plants. Proc Natl Acad Sci 111:6497–6502. https://doi.org/10.1073/pnas.1319955111
8. Szechyńska-Hebda M, Kruk J, Górecka M et al (2010) Evidence for light wavelength-specific photoelectrophysiological signaling and memory of excess light episodes in Arabidopsis. Plant Cell 22:2201–2218. https://doi.org/10.1105/tpc.109.069302
9. Koziolek C, Grams TEE, Schreiber U et al (2004) Transient knockout of photosynthesis mediated by electrical signals. New Phytol 161:715–722. https://doi.org/10.1111/j.1469-8137.2004.00985.x
10. Hlaváčková V, Krchňák P, Nauš J et al (2006) Electrical and chemical signals involved in short-term systemic photosynthetic responses of tobacco plants to local burning. Planta 225:235–244. https://doi.org/10.1007/s00425-006-0325-x
11. Kaiser H, Grams TEE (2006) Rapid hydropassive opening and subsequent active stomatal closure follow heat-induced electrical signals in Mimosa pudica. J Exp Bot 57:2087–2092. https://doi.org/10.1093/jxb/erj165
12. Grams TEE, Lautner S, Felle HH et al (2009) Heat-induced electrical signals affect cytoplasmic and apoplastic pH as well as photosynthesis during propagation through the maize leaf. Plant Cell Environ 32:319–326. https://doi.org/10.1111/j.1365-3040.2008.01922.x
13. Gallé A, Lautner S, Flexas J et al (2013) Photosynthetic responses of soybean (Glycine max L.) to heat-induced electrical signalling are predominantly governed by modifications of mesophyll conductance for CO_2. Plant Cell Environ 36:542–552. https://doi.org/10.1111/j.1365-3040.2012.02594.x
14. Devireddy AR, Zandalinas SI, Gómez-Cadenas A et al (2018) Coordinating the overall stomatal response of plants: rapid leaf-to-leaf communication during light stress. Sci Signal 11:eaam9514. https://doi.org/10.1126/scisignal.aam9514
15. Devireddy AR, Arbogast J, Mittler R (2020) Coordinated and rapid whole-plant systemic stomatal responses. New Phytol 225:21–25. https://doi.org/10.1111/nph.16143
16. Ehonen S, Hölttä T, Kangasjärvi J (2020) Systemic signaling in the regulation of stomatal

conductance. Plant Physiol 182:1829–1832. https://doi.org/10.1104/pp.19.01543

17. Zandalinas SI, Cohen IH, Fritschi FB, Mittler R (2020) Coordinated systemic stomatal responses in soybean. Plant Physiol 183: 1428–1431. https://doi.org/10.1104/pp.20.00511
18. Hõrak H, Fountain L, Dunn JA et al (2020) Dynamic thermal imaging confirms local but not fast systemic ABA responses. Plant Cell Environ 44:885–899. https://doi.org/10.1111/pce.13973
19. Farmer E, Farmer E, Mousavi S, Lenglet A (2013) Leaf numbering for experiments on long distance signalling in Arabidopsis. Protoc Exch. https://doi.org/10.1038/protex.2013.071
20. Vahisalu T, Puzõrjova I, Brosché M et al (2010) Ozone-triggered rapid stomatal response involves the production of reactive oxygen species, and is controlled by SLAC1 and OST1. Plant J 62:442–453. https://doi.org/10.1111/j.1365-313X.2010.04159.x
21. Zuur AF, Ieno EN, Elphick CS (2010) A protocol for data exploration to avoid common statistical problems. Methods Ecol Evol 1:3–14. https://doi.org/10.1111/j.2041-210x.2009.00001.x
22. Wood SN (2006) Generalized additive models: an introduction with R, 2nd edn. Chapman & Hall/CRC, Boca Raton, FL
23. Rose NL, Yang H, Turner SD, Simpson GL (2012) An assessment of the mechanisms for the transfer of lead and mercury from atmospherically contaminated organic soils to lake sediments with particular reference to Scotland, UK. Geochim Cosmochim Acta 82: 113–135. https://doi.org/10.1016/J.GCA.2010.12.026

Chapter 10

Liquid Chromatography–Tandem Mass Spectrometry-Based Profiling of Plant Hormones

Mohamed A. Salem and Ahmed Zayed

Abstract

Phytohormones plays crucial physiological functions in plants, where they are involved in plant development, reproduction, defense, and many other functions. Phytohormones production has been found to be regulated in response to abiotic and biotic factors affecting the plant metabolism, and therefore, biosynthesis of primary and secondary metabolites. Thus, the detection and quantification of phytohormones in different plant tissues are essential to be determined unraveling the various plant metabolic pathways and behavior. Yet phytohormones analysis is always problematic, since they are found in extremely low concentrations and have a wide range of chemical and physicochemical properties. As a result, the ideal method should start with an appropriate extraction procedure followed by quantification by highly sensitive instrumental techniques providing precise and robust results. The current chapter presents an improved extraction method based on liquid-liquid extraction from a 50-mg aliquot of plant tissue for analysis of the major classes of phytohormones. Then, mass spectrometry (MS) analysis is conducted using quadrupole/linear ion trap (QLIT) mass analyzer equipped with electrospray ionization (ESI) source after a liquid chromatographic separation step. The developed method demonstrates an appropriate feasibility addressing biological questions related to phytohormones production and regulation.

Key words Extraction, High resolution mass spectrometry, Liquid chromatography, Plant hormones, Quantitative analysis

1 Introduction

Stress conditions either biotic or abiotic encourage plants developing various adaptation strategies, including regulation of hormonal levels triggering physiological, biochemical, and molecular responses, which affect growth and development processes [1]. Plant hormones, or phytohormones, have crucial potential physiological roles in response to biotic and abiotic stresses [2]. Plant hormones are involved in cell division, development, reproduction, and defense [3, 4]. With chemically distinguishable nine classes, including auxins (Aux), jasmonates (JAs), gibberellins (GAs), strigolactones (STs), salicylic acid (SA), abscisic acid (ABA),

Takuya Yoshida (ed.), *Abscisic Acid: Methods and Protocols*,
Methods in Molecular Biology, vol. 2462, https://doi.org/10.1007/978-1-0716-2156-1_10,

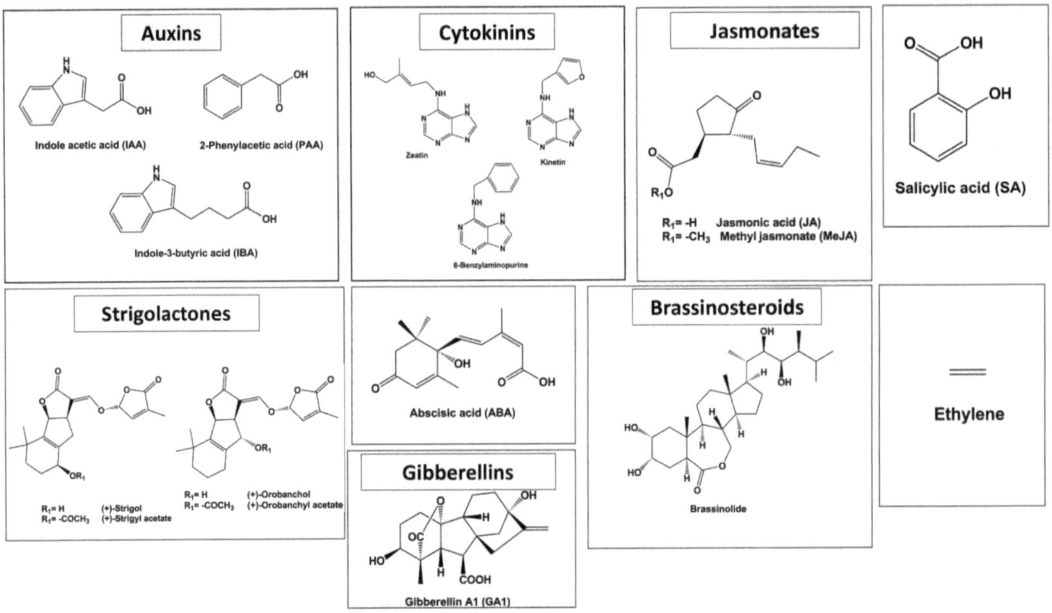

Fig. 1 Classes of phytohormones with the chemical structure of some representatives

cytokinins (CKs), brassinosteroids (BRs), and ethylene (ET), the plant various signaling systems, metabolites transport, and metabolism are coordinated [5–7]. Recently, the advances in omics technology have proven the interaction between plant hormones and biological processes [8]. The chemical structure of some examples representing the nine classes are depicted in Fig. 1.

Particularly, ABA serves in seeds dormancy, bud growth, and plant adaptation, especially in perennial trees during autumn and winter facing the extreme environmental conditions [9]. In addition, previous reports have shown that ABA level is increased under drought stress condition to act as a key signaling molecule inducing stomatal closure and transcriptional changes in leaves [10, 11].

Yet, recent investigations have revealed that the phytohormones profile, including jasmonates, SA, and cytokinin, that is, *trans*-zeatin is dynamically changed also in response to drought stress [2, 12, 13]. Therefore, simultaneous multiple phytohormones quantification should be taken into consideration compared with, for example, sole ABA measurement. These findings have proven the complex plant interactions with its environment.

However, multiple phytohormone screening is challenging, since their concentrations are in fmol to pmol/g fresh weight of plant tissues, compared to primary and secondary metabolites [14]. Accordingly, an optimized analytical method should provide a qualitative overview and precise quantitative information facilitating the study of phytohormones signaling networks. These objectives require a convenient sample preparation followed by high instrumental performance guaranteeing robustness and sensitivity [14]. Separation by liquid chromatography (LC) or gas chromatography (GC) coupled with tandem mass spectrometric detection (MS/MS), especially electrospray ionization (ESI), are the most widely applied methods [2, 14, 15].

The current chapter presents an optimized, easy, and valid extraction method followed by quantitative determination using both chromatographic and spectral data for the multiple phytohormone profiling from a 50 mg aliquot of plant tissue based on our recently developed protocol [2]. The aliquot of the extracted hormones is subjected to separation by reversed-phase ultraperformance LC (RP-UPLC) coupled to ESI and a quadrupole/linear ion trap (QLIT) mass analyzer.

2 Materials

Use solvents of LC grade. Perform all the extraction steps in a fume hood cabinet. Wear protective glasses and clothing during the extraction.

2.1 Reagents and Chemicals

1. Authentic plant hormones: Abscisic acid, ABA; Gibberellic acid, GA3; Gibberellin A4, GA4; Indole-3-acetic acid, IAA; Jasmonic acid, JA; Jasmonoyl-L-isoleucine, JA-Ile; Salicylic acid, SA; *trans*-Zeatin (tZ); 12-oxo-Phytodienoic acid, OPDA (*see* **Note 1**).
2. Internal standards (IS) ([$^2H_6$](+) *cis*, *trans*-Abscisic acid, [$^2H_2$]-Gibberellin A4, Indol-3-acetic-2,2-[$^2H_5$] acid, [$^2H_2$] (±)-Jasmonic Acid, [$^2H_3$]*N*-[(-)-Jasmonoyl]-Leucine, [$^2H_5$] *cis*-12-Oxo-Phytodienoic Acid, [$^2H_4$] Salicylic acid, [$^2H_5$] *trans*-Zeatin) (*see* **Note 1**).

2.2 Plant Material Harvesting and Storage

1. Harvesting bottles or tubes.
2. Scissors and tweezers.
3. Mortar and pestle or tissue grinder.
4. Liquid nitrogen.
5. −80 °C freezer.
6. Analytical balance.

2.3 Hormone Extraction

1. Standard solutions: Dissolve the individual standards in LC-grade methanol (MeOH) at a stock of 1 mg/mL (*see* **Notes 2** and **3**).
2. Extraction solvent: Prepare a mixture of methyl *tert*-butyl ether (MTBE) and MeOH (3:1 v/v) (*see* **Note 4**).

3. Solvent to induce the phase separation: Prepare 0.1% (v/v) HCl in water.
4. Solvent for re-suspension for the LC-MS measurement: Prepare 50% (v/v) MeOH in water.
5. Solvent used to obtain the MRM transitions for authentic hormones and IS: 50% MeOH, 0.1% formic acid in water.
6. Microcentrifuge tubes or preferably glass vials.
7. Vortex mixer.
8. Sonicator.
9. Benchtop centrifuge.
10. Nitrogen (N_2) dryer or vacuum concentrator.
11. Orbital shaker.

2.4 Liquid Chromatography and Mass Spectrometry

1. LC system (e.g., Waters Acquity UPLC system or equivalent), with auto-sample, column oven and binary pumps.
2. Reversed phase (RP) C18-column (high strength silica, HSS, T3 column, 100 mm × 2.1 mm containing 1.8 μm diameter particles) (e.g., Waters or equivalent) fitted with 2.1 × 5 mm guard column (ACQUITY UPLC HSS T3 VanGuard Pre-column, 100 Å, 1.8 μm) (e.g., Waters or equivalent).
3. LC Solvent A: 0.1% v/v formic acid in water.
4. LC Solvent B: 0.1% v/v formic acid in MeOH.
5. Autosampler glass vials with inserts.
6. MS instrument (e.g., quadrupole/linear ion trap (QLIT) mass analyzer, i.e., AB Sciex 4000 QTRAP or equivalent), equipped with electrospray ionization (ESI) source (e.g., AB Sciex Turbo V™ Ion Source or equivalent), coupled with LC system.
7. Software (e.g., AB Sciex Analyst software, version 1.6.2 or equivalent).

3 Methods

3.1 Tissue Harvesting, Quenching, and Homogenization

1. Harvest the plant tissue and snap-freeze the samples in liquid nitrogen (*see* **Note 5**).
2. Homogenize the plant tissue using a mortar and pestle or tissue homogenizer (*see* **Note 6**).
3. Store samples in liquid N_2, dry ice or −80 °C freezer until further extraction (*see* **Note 7**).

3.2 Extraction of Hormone Samples

1. Weigh 50 mg of the ground plant tissue in 2 mL microcentrifuge tubes or preferably glass vials. Record the sample weight (*see* **Note 6**).

2. Add 1 mL of precooled (−20 °C) extraction solvent to each sample (*see* **Note 8**).
3. Vortex samples until all tissue is properly suspended.
4. Keep samples on an orbital shaker for 45 min at 4 °C at 100 rpm.
5. Put the samples on an ice-cooled sonicator for 15 min.
6. Add 0.5 mL of solvent used for phase separation.
7. Vortex the samples for 1 min.
8. Centrifuge the samples at 10,000 × *g* for 10 min at 4 °C.
9. Transfer precisely 0.8 mL of the upper phase to in microcentrifuge tubes or preferably glass vials.
10. Dry the samples under N_2 gas or in a vacuum concentrator without heating.
11. Add 0.1 mL of resuspension solvent, vortex the samples and centrifuge at 5000 × *g* for 3 min at 4 °C.
12. Transfer the samples to autosampler glass vials and cap tightly.

3.3 LC-MS Analysis

1. Set up the LC separation method. Set the autosampler temperature at 10 °C.
2. Set the column temperature to 40 °C.
3. Set the flow rate at 0.4 mL/min.
4. Inject 7.5 μL per sample.
5. Start the LC gradient in negative ionization mode with 62% solvent A for 6.5 min; 45% solvent B from 6.5 to 7 min; 10% solvent A from 7 to 7.1 min, held at 0% solvent A from 7.1 to 8.1 min, returned to initial conditions by 8.1 min. From 8.1 to 10 min, reequilibrate and condition the column to 62% solvent A. Perform the analysis in the positive mode using the same gradient but starting with 80% eluent A instead of 62% eluent A.
6. Acquire mass spectra from LC runs using a suitable quadrupole/linear ion trap (QLIT) mass analyzer in multiple reaction monitoring (MRM) scan type equipped with an electrospray ionization (ESI) source.
7. Obtain the MRM transitions for authentic hormones and IS by direct infusion of each compound into the mass spectrometer at 10 μL/min (*see* **Note 9**).
8. Adjust the electronic settings for the mass spectrometer as following: ion polarity; positive or negative, ion source; Turbo spray, capillary ion spray voltage; 5.5 kV or −4.5 kV for positive or negative ion polarity, respectively, probe tip position; 3 mm, source temperature; 500 °C, gas; nitrogen,

curtain gas; 25 psi, nebulizing gas (GS1) and focusing gas (GS2); 40 psi, interface heater; On, collision activated dissociation gas pressure; Medium.

9. Use a suitable software from the vendor for instrument control, data acquisition, processing, and analysis.

3.4 Hormone Identification and Quantification

1. Identify hormones using the chromatographic and spectral data (Table 1).
2. Perform the quantification using external calibration standards.
3. Analyze the standard mixtures containing all phytohormones at different concentrations (e.g., 1000, 100, 10, 1, 0.1, and 0.01 ng/mL).

Table 1
Expected retention time, molecular ion peaks, and characteristic fragments of some selected phytohormones

Phytohormone	Expected retention time (min)	Ionization mode	Q1 precursor ion (*m/z*)	Q3 product ions (*m/z*)	Quantitation peak (*m/z*)
trans-Zeatin (tZ)	1.46	Positive	220.15 for $[M + H]^+$	135.9, 202	135.9
Gibberellic acid (GA_3)	1.87	Negative	345.09 for $[M - H]^-$	142.9, 239	142.9
Indole-3-carboxylic acid (ICA)	2.22	Negative	160.001 for $[M - H]^-$	115.9, 65.8	115.9
Salicylic acid (SA)	3.08	Negative	136.698 for $[M - H]^-$	64.9, 92.8	92.8
Abscisic acid (ABA)	4.13	Negative	263.04 for $[M - H]^-$	152.9, 219	152.9
Indole-3-acetic acid (IAA)	4.67	Positive	176.118 for $[M + H]^+$	130, 77.1	130
Indole-3-butyric acid (IBA)	4.9	Negative	202.017 for $[M - H]^-$	157.9, 115.7	157.9
Jasmonic acid (JA)	5.58	Negative	209.072 for $[M - H]^-$	58.9, 96.8	58.9
Gibberellin A_4 (GA_4)	7.78	Positive	333.215 for $[M + H]^+$	315.1, 269.2	315.1
Brassinolide (BL)	7.85	Positive	481.422 for $[M + H]^+$	95, 71	95
12-oxo-Phytodienoic acid (OPDA)	8.0	Positive	293.26 for $[M + H]^+$	275.2, 79	275.2

4. Plot the peak areas of hormones against their concentration.
5. Fit the curves by least-squares linear regression using 1/× weighting factor. Get the regression equation and correlation coefficient (R^2) from the vendor software.

4 Notes

1. Store the standard hormones according to the recommendations of the supplier.
2. Store the stock solution protected from light at −20 °C until use.
3. Working solutions of authentic hormones (ranging from 1000 to 0.01 ng/mL) and IS (50 ng/mL) prepared by serial dilution of the stock solution in methanol/water (50:50 v/v) prior to analysis.
4. Extraction solvent mixture contained MTBE: MeOH (3:1 v/v) is prepared together with isotope-labeled IS at a concentration of 50 ng/mL and the extraction solvent is kept at −20 °C [16, 17].
5. The harvesting and freezing step has to be performed quickly to avoid changes due to enzymatic activities.
6. Thawing of the plant tissues should be avoided during grinding or weighing steps, to minimize enzymatic degradation.
7. Samples should be extracted directly after grinding and long-term storage should be avoided.
8. To strengthen the quantitative character of the method, use of isotope-labeled IS for each hormone for the compensation of losses during sample preparation or analysis. These isotope-labeled IS have similar extraction efficiencies and chromatographic behavior as the corresponding nonlabeled compounds [18].
9. Direct infusion of each compound in the mass spectrometer for obtaining the MRM transitions requires dissolving the appropriate standard in 50% MeOH/water containing 0.1% formic acid at a concentration of 100 ng/mL. A schematic diagram (Fig. 2) is illustrated to summarize the protocol steps used for sample extraction and analysis.

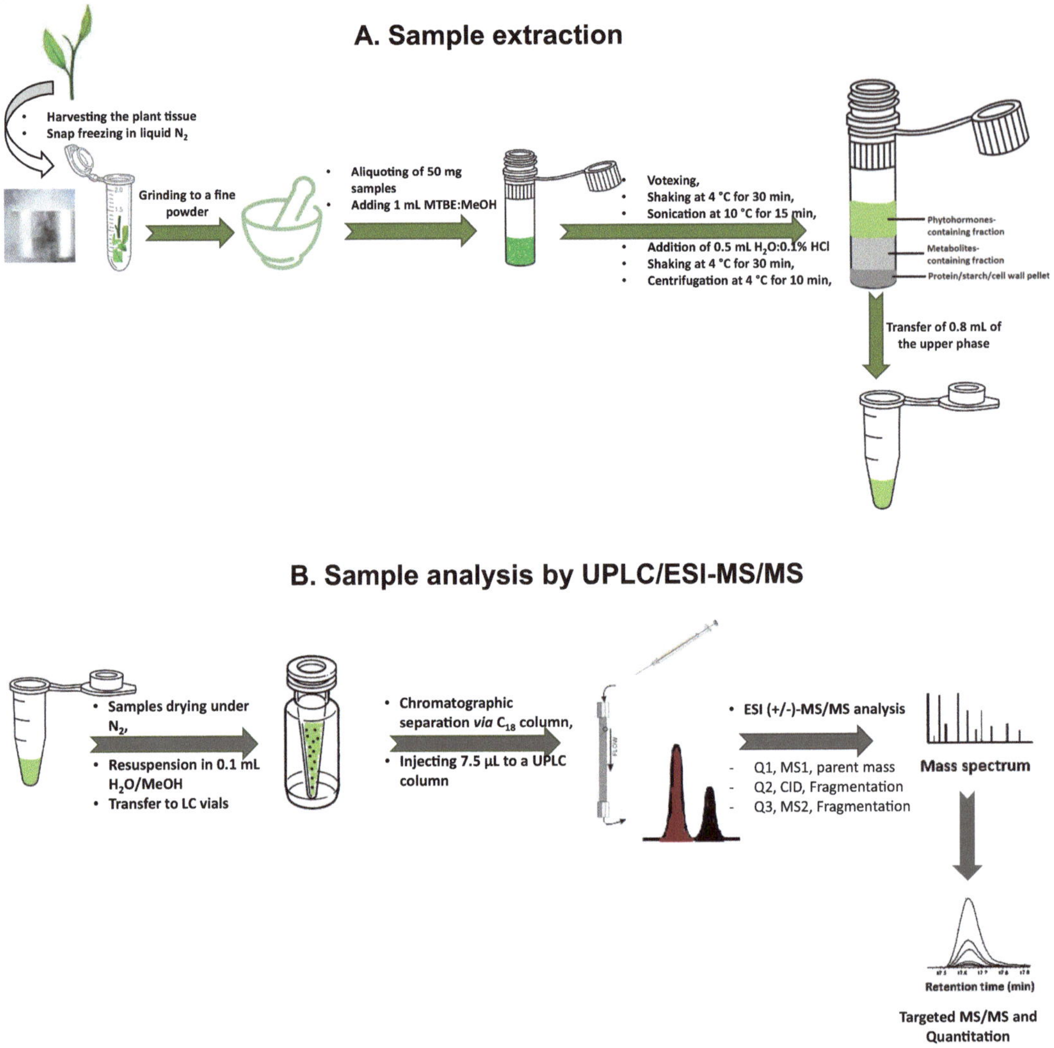

Fig. 2 (**a**) A schematic diagram summarizes the phytohormone extraction; (**b**) Analysis protocol by UPLC/ESI-MS/MS

References

1. Sah SK, Reddy KR, Li J (2016) Abscisic acid and abiotic stress tolerance in crop plants. Front Plant Sci 7:571
2. Salem MA, Yoshida T, Perez de Souza L, Alseekh S, Bajdzienko K, Fernie AR, Giavalisco P (2020) An improved extraction method enables the comprehensive analysis of lipids, proteins, metabolites and phytohormones from a single sample of leaf tissue under water-deficit stress. Plant J 103(4):1614–1632
3. Kim SW, Goossens A, Libert C, Van Immerseel F, Staal J, Beyaert R (2020) Phytohormones: multifunctional nutraceuticals against metabolic syndrome and comorbid diseases. Biochem Pharmacol 175:113866
4. Skalický V, Kubeš M, Napier R, Novák O (2018) Auxins and cytokinins-The role of subcellular organization on homeostasis. Int J Mol Sci 19(10):3115
5. Larrieu A, Vernoux T (2015) Comparison of plant hormone signalling systems. Essays Biochem 58:165–181
6. Mishra S, Upadhyay S, Shukla RK (2017) The role of strigolactones and their potential cross-

talk under hostile ecological conditions in plants. Front Physiol 7:691

7. Sharma I, Kaur N, Pati PK (2017) Brassinosteroids: a promising option in deciphering remedial strategies for abiotic stress tolerance in rice. Front Plant Sci 8:2151
8. Nawaz I, Tariq R, Nazir T, Khan I, Basit A, Gul H, Anwar T, Awan SA, Bacha SAS, Zhang L, Zhang C, Cong P (2021) RNA-Seq profiling reveals the plant hormones and molecular mechanisms stimulating the early ripening in apple. Genomics 113(1 Pt 2): 493–502
9. Wang D, Gao Z, Du P, Xiao W, Tan Q, Chen X, Li L, Gao D (2016) Expression of ABA metabolism-related genes suggests similarities and differences between seed dormancy and bud dormancy of peach (Prunus persica). Front Plant Sci 6:1248
10. Ali S, Hayat K, Iqbal A, Xie L (2020) Implications of abscisic acid in the drought stress tolerance of plants. Agronomy 10(9):1323
11. Yoshida T, Christmann A, Yamaguchi-Shinozaki K, Grill E, Fernie AR (2019) Revisiting the basal role of ABA - roles outside of stress. Trends Plant Sci 24(7):625–635
12. Gupta A, Hisano H, Hojo Y, Matsuura T, Ikeda Y, Mori IC, Senthil-Kumar M (2017) Global profiling of phytohormone dynamics during combined drought and pathogen stress in Arabidopsis thaliana reveals ABA and JA as major regulators. Sci Rep 7(1):4017
13. Urano K, Maruyama K, Jikumaru Y, Kamiya Y, Yamaguchi-Shinozaki K, Shinozaki K (2017) Analysis of plant hormone profiles in response to moderate dehydration stress. Plant J 90(1): 17–36
14. Šimura J, Antoniadi I, Široká J, Tarkowská D, Strnad M, Ljung K, Novák O (2018) Plant hormonomics: multiple phytohormone profiling by targeted metabolomics. Plant Physiol 177(2):476–489
15. Cai W-J, Ye T-T, Wang Q, Cai B-D, Feng Y-Q (2016) A rapid approach to investigate spatiotemporal distribution of phytohormones in rice. Plant Methods 12(1):1–10
16. Salem MA, Jüppner J, Bajdzienko K, Giavalisco P (2016) Protocol: a fast, comprehensive and reproducible one-step extraction method for the rapid preparation of polar and semi-polar metabolites, lipids, proteins, starch and cell wall polymers from a single sample. Plant Methods 12(1):45
17. Salem M, Bernach M, Bajdzienko K, Giavalisco P (2017) A simple fractionated extraction method for the comprehensive analysis of metabolites, lipids, and proteins from a single sample. J Vis Exp 124:55802
18. Stokvis E, Rosing H, Beijnen JH (2005) Stable isotopically labeled internal standards in quantitative bioanalysis using liquid chromatography/mass spectrometry: necessity or not? Rapid Commun Mass Spectrom 19(3): 401–407

Chapter 11

Live Imaging of Abscisic Acid Dynamics Using Genetically Encoded Fluorescence Resonance Energy Transfer (FRET)-Based ABA Biosensors

Rainer Waadt

Abstract

The phytohormone abscisic acid (ABA) regulates various aspects of plant physiology, growth, and development to maintain a balanced plant water status. Cellular ABA levels are regulated through the combined activities of biosynthesis, catabolism, and transport proteins and depend on the developmental stage, the cell-type and on environmental conditions. Genetically encoded Förster (fluorescence) Resonance Energy Transfer (FRET)-based ABA-responsive biosensors enable the direct monitoring of ABA dynamics in intact plants. Thus, ABA biosensor-based in vivo imaging can provide novel insights about the spatiotemporal patterns of biosynthesis- and transport-dependent ABA dynamics that are required for the regulation of seed dormancy and germination, root growth and hydrotropism, and stomatal closure under water limiting conditions. Here, I describe a protocol for the in vivo analysis of ABA in 5-day-old Arabidopsis seedlings (roots) expressing the FRET-based ABA biosensor ABAleonSD1-3L21.

Key words ABAleon, Abscisic acid, *Arabidopsis thaliana*, Biosensor, FRET, Live-imaging

1 Introduction

Live imaging in plants using genetically encoded FRET-based biosensors has been initially employed to report on ABA-induced Ca^{2+} dynamics in Arabidopsis guard cells two decades ago [1]. Since then, many biological findings have been made using FRET-based and other fluorescent biosensors [2–4]. The direct in vivo analysis of cytosolic ABA levels has been made possible through the development of ABACUS1- and ABAleon-type FRET-based biosensors (Table 1) [5, 6]. These biosensors consist of an ABA-responsive sensory module harboring the full-length sequence of either PYRABACTIN RESIS TANCE 1 (PYR1, also known as REGULATORY COMPONENT OF ABA RECEPTOR 11 (RCAR11)) or PYR1-LIKE 1 (PYL1, also known as RCAR12) ABA receptors [8, 9] fused via a linker to a fragment of the coreceptor type 2C protein phosphatase (PP2C)

Takuya Yoshida (ed.), *Abscisic Acid: Methods and Protocols*,
Methods in Molecular Biology, vol. 2462, https://doi.org/10.1007/978-1-0716-2156-1_11,

Table 1
Properties of FRET-based ABA-responsive biosensors. Shown are the biosensor configuration, ABA dissociation constants (K_d), the maximum in vitro emission ratio change ($\Delta R/R_0$), the in vivo emission ratio change in roots in response to 60 min treatments with 10 μM ABA and the associated half-response time ($t_{1/2}$) of each FRET-based ABA biosensor. The data were taken from the references indicated in brackets

FRET-based ABA biosensor	Biosensor configuration	In vitro ABA dissociation constant (K_d [nM])	Maximum ABA-dependent emission ratio change ($\Delta R/R_0$) in vitro	60 min 10 μM ABA-dependent emission ratio change ($\Delta R/R_0$) in roots	10 μM ABA half-response time in roots ($t_{1/2}$ [min])
ABACUS1-80μ	edCitrine (acceptor)-B1 linker-ABI1(H279-D327)-52 aa spring linker-PYL1-B2 linker-edCerulean (donor)	77000 ± 5300 [5]	+1.545 [5]	Not determined	Not determined
ABACUS1-2μ	ABACUS1-80μ with PYL1 H87P mutation	1800 ± 90 [5]	+0.938 [5]	+0.182 [7]	29 [7]
ABACUS1-nr	ABACUS1-80μ with PYL1 K86A mutation	nonresponsive control [5]	± 0 [5]	Not determined	Not determined
ABAleon2.1	mTurquoise (donor)-GP linker-PYR1-29 aa GS linker-ABI1(S125-N434, D413L mutation)-PG linker-cpVenus173 (acceptor)	114 ± 32 [6]	−0.090 [6]	−0.11 [6]	16 [6]
ABAleon2.15	ABAleon2.1 with PYR1 H115A mutation	510 ± 41 [6] 817 ± 32 [7]	−0.101 [6] −0.185 [7]	−0.08 [6] −0.128 [7]	12 [6] 16 [7]
ABAleonSD1-3L21	PmTurquoise (donor)-LD linker-PYR1(S3-T191, H115A mutaton)-29 aa GS linker-ABI1(S125-N434, D413L mutation)-T linker-cpVenus173 (acceptor)	938 ± 22 [7]	−0.167 [7]	−0.138 [7]	15 [7]

ABI1, which is critical for PYR1/PYL1-ABI1-mediated ABA binding [10–12]. The ABA sensory module is further flanked by a fluorescent protein FRET-pair, with a cyan fluorescent protein (CFP) variant functioning as FRET donor and a yellow fluorescent protein (YFP) variant functioning as FRET acceptor. The configuration of ABACUS1 and ABAleon biosensors [5–7] is summarized in Table 1. ABA receptors and PP2C coreceptors belong to large gene families with many potential combinations that could function as a sensory module in a FRET-based reporter. Therefore, the ABA-responsive sensory modules in ABACUS1-2μ and ABAleon2.1 were designed on the basis of previous structural analyses [10–12] and by screening several sensory module configurations harboring various ABA receptors or fragments of receptor and PP2C coreceptor pairs [5, 6].

There are two requirements for FRET to occur. (1) The emission spectrum of the donor fluorophore must overlap with the excitation spectrum of the acceptor fluorophore, and (2) both donor and acceptor fluorophores must be in close 1–10 nm proximity with an appropriate orientation of their dipole vectors [13, 14]. The general readout of a FRET-based biosensor is the emission ratio of the acceptor over the donor upon excitation of the donor (i.e., YFP/CFP emission ratio upon CFP excitation). With increasing ABA concentrations, the emission ratio of ABACUS1 is increasing, whereas the ABAleon emission ratio is decreasing [5–7] (Table 1). In live-imaging approaches, such ABA-dependent emission ratio changes can be assessed through fluorescence microscopy analyses and thus allow for spatiotemporal analyses of cytosolic ABA concentration changes in vivo.

Because both ABACUS1- and ABAleon sensory modules consist of PYR1 or PYL1 and ABI1 fragments, the expression of ABACUS1-2μ in Arabidopsis leads to ABA hypersensitivity effects, whereas ABAleon expression renders Arabidopsis plants ABA hyposensitive [5–7, 15]. In ABACUS1-2μ, the ABI1 fragment is very small (49 amino acids) and likely has no side effects on ABA signaling. However, the incorporated PYL1 component with the H87P mutation, that renders PYL1 from a dimeric to a monomeric ABA receptor, exhibits an increased probability to interact with other cellular PP2C coreceptor proteins in the absence of ABA, thus shifting the ABA signaling equilibrium toward an increased ABA sensitivity [8, 9, 16]. In ABAleons, the ABA sensory module consists of PYR1 and the entire ABI1 phosphatase domain [6, 7]. PYR1 is dimeric and interacts with PP2Cs only in the presence of ABA [9, 16, 17]. Because the ABAleon PYR1 component is directly linked to the ABI1 phosphatase domain, ABA-dependent intramolecular interactions are more likely. However, in the absence of ABA, it is possible that the ABI1 phosphatase domain component, although rendered inactive through a catalytic site D413L mutation, interacts with and modulates the function of cellular target proteins. It is also possible that ABAleons buffer ABA at physiologically relevant concentrations. Therefore, the effects of these ABA biosensors on ABA signaling should be considered when interpreting the obtained live-imaging results.

Initially, ABACUS1-2µ and ABAleon2.1 were used for the analysis of ABA uptake and transport in Arabidopsis guard cells, roots, and hypocotyls. These biosensors also successfully reported slow (4–24 h) salt- and osmotic stress-induced ABA accumulation in Arabidopsis guard cells, leaves and roots, and 15 min low humidity-induced ABA accumulation in Arabidopsis guard cells [5, 6]. Later research using ABAleons, revealed that 2 h treatments with sulfate or cysteine trigger ABA accumulation in Arabidopsis guard cells [18], whereas elevated CO_2 had no rapid effect on guard cell ABA levels [19, 20]. Simultaneous analyses of ABAleons together with the red-fluorescing Ca^{2+} indicator R-GECO1 [21] or the pH indicator (P)A-17 [22] revealed that ABA uptake in Arabidopsis roots had no rapid effect on cytosolic Ca^{2+} and H^+ levels [7, 23]. Note that environmentally triggered ABA elevations in Arabidopsis roots might take several hours until leaf-synthesized ABA is transported to the roots [6, 24–26]. The spatiotemporal parameters of environmentally triggered ABA biosynthesis, catabolism, and transport in leaves (vascular tissues and guard cells) [27–29] and during seed development and germination [30] also remain to be determined. Analyses using FRET-based ABA biosensors could provide novel insights in this respect.

Several previously published protocols describe FRET-based biosensor live-imaging analyses in Arabidopsis [31–35]. The below protocol has been developed and refined during the course of three publications [6, 7, 23] and describes a basic ABA uptake experiment in 5-day-old Arabidopsis seedling roots using the FRET-based ABA biosensor ABAleonSD1-3L21. Such or similar analyses might be useful for the characterization of ABA transporter mutants in Arabidopsis. This protocol can be also easily adapted for time-lapse in vivo imaging analyses in Arabidopsis roots expressing any type of biosensor [7]. Along with the experimental workflow described below, it is also recommended to conduct additional mock-treatment controls for better interpretation of the observed results.

2 Materials

Media and buffers are prepared using ultrapure water (e.g., Milli-Q).

2.1 Plant Material

Arabidopsis seeds expressing FRET-based ABA biosensors (Table 1) [5–7] (*see* **Note 1**).

2.2 Plant Growth Medium (LAK) [36]

1. Macro 1 (100×): 100 mM KH_2PO_4. Sterilize by autoclaving and store at 4 °C.
2. Macro 2 (100×): 200 mM $Ca(NO_3)_2 \cdot 4H_2O$. Sterilize by autoclaving and store at 4 °C.
3. Macro 3 (100×): 100 mM $MgSO_4 \cdot 7H_2O$. Sterilize by autoclaving and store at 4 °C.
4. Micros (1000×): 30 mM H_3BO_3, 10 mM $MnSO_4 \cdot 1H_2O$, 10 mM $ZnSO_4 \cdot 7H_2O$, 1 mM $CuSO_4 \cdot 5H_2O$, 0.03 mM $(NH_4)_6Mo_7O_{24} \cdot 4H_2O$. Sterilize by autoclaving and store at 4 °C.
5. Ethylenediaminetetraacetic acid ferric-sodium salt (FeNaEDTA; 50 mM; 1000×). Sterilize by autoclaving and store at 4 °C.
6. LAK Medium: For 1 L medium add under sterile conditions 10 mL of Macro 1, 10 mL of Macro 2, 10 mL of Macro 3, 1 mL of Micros, and 1 mL of 50 mM FeNaEDTA. Supplement the solution with 1.95 g 2-(*N*-Morpholino)ethanesulfonic acid (MES; MW 195.24) and adjust pH to 5.6 using Tris Base powder. Transfer the solution to a 1 L bottle containing 0.8% (w/v) Phytoagar and autoclave. After autoclaving cool down the solution to ca. 60 °C while steering. Under sterile conditions, pour the media (ca. 67 mL) into 12 cm diameter square petri dishes. After polymerization (at least 1 h), the petri dishes can be stored in a plastic bag for 1 month at 4 °C (*see* **Note 2**).

2.3 Media for Microscopy

1. Microscope Dish Medium: For 100 mL medium add under sterile conditions 1 mL of Macro 1, 1 mL of Macro 2, 1 mL of Macro 3, 100 μL of Micros, and 100 μL of 50 mM FeNaEDTA. Supplement the solution with 195 mg MES and adjust pH to 5.6 using Tris Base powder. Transfer the solution to a 100 mL bottle containing 0.7% (w/v) low melting point agarose, and autoclave. After autoclaving, allow polymerization of the solution at room temperature (RT). To avoid nonhomogeneous polymerization, warm up the medium in a microwave, let it cool down at RT and store it at 4 °C (*see* **Note 2**).
2. Plant Imaging Buffer [37]: 5 mM KCl, 50 μM $CaCl_2$ diluted from 1 M stock solution, 10 mM MES, adjust pH to 5.6 using Tris Base powder. Prepare 1 L solution, sterilize by filtration through a 0.45 μm filter and store at RT. For microscopy experiments, transfer under sterile conditions an aliquot into a 50 mL tube.
3. ABA Stock Solution: 100 mM (*S*)-(+)-abscisic acid dissolved in 100% EtOH. Weight a 1.5 mL tube on a fine balance and add ca. 5 mg ABA into the tube. Calculate the exactly required

amount of EtOH and add this amount to the tube. Mix thoroughly by pipetting. Tightly close the tube with Parafilm and store it at −20 °C.

2.4 Material for Seed Sterilization and Sowing

1. Overhead rotating device for 2 mL reaction tubes.
2. Laminar hood.
3. 70% EtOH.
4. 100% EtOH.
5. 2 mL reaction tubes.
6. Sterile paper.
7. Sterile toothpicks.
8. Micropore tape.
9. Aluminum foil.

2.5 Material for Plant Growth and Selection

1. Plant Growth Cabinet: 16 h day/8 h night cycle, 22 °C, 65% relative humidity and 150 μmol/m^2/s photon fluence rate.
2. Microscope (glass-bottom) dishes: dish diameter 50 mm, glass-bottom well diameter 14 mm, glass thickness No. 0 or No. 1.5 (*see* **Note 3**).
3. Forceps for transferring seedlings to microscope dishes.
4. Holders that enable the incubation of square petri dishes in a vertical position.
5. Holders that enable the incubation of microscope dishes in a vertical position.
6. Fluorescence stereomicroscope equipped with a 1× objective and a GFP or YFP filter set (*see* **Note 4**).

2.6 Material for Microscopy, Image Processing, and Data Analysis

1. Inverted Leica SP8 confocal laser scanning microscope equipped with a 10× objective, a 458 nm laser, photomultiplier tube detectors and suitable for microscope dish mounting (*see* **Note 5**).
2. Fiji ImageJ [38].
3. GEFI_Analyzer plugin for Fiji [7].
4. Microsoft Excel.
5. OriginLab OriginPro or other graphing software.
6. Adobe Illustrator, or other graphic design tool.

3 Methods

A schematic workflow of the entire experiment is depicted in Fig. 1.

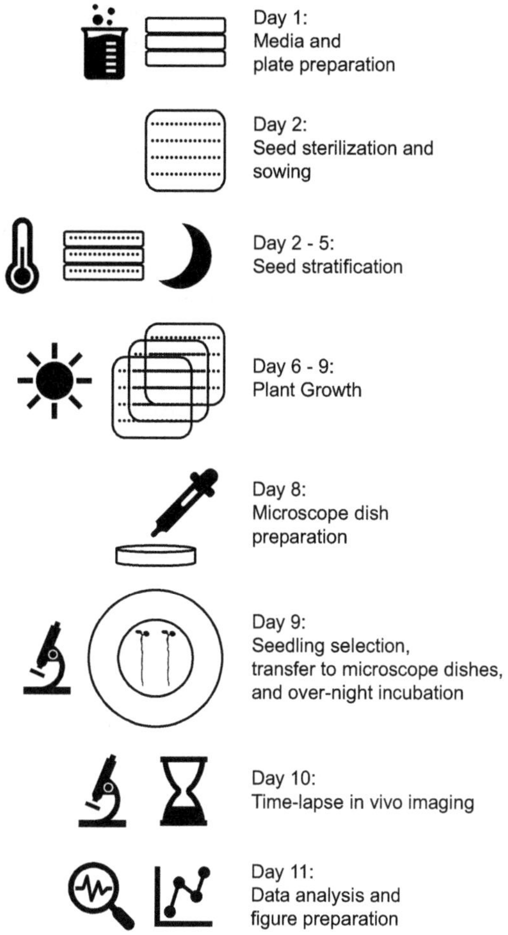

Fig. 1 Schematic overview of a biosensor-based in vivo imaging experiment in Arabidopsis roots with indicated daily major tasks

3.1 Seed Sterilization, Sowing and Plant Growth

1. Add ca. 30 μL of Arabidopsis ABA biosensor seeds into a 2 mL tube.
2. Under a laminar hood, add 1 mL of 70% EtOH to the tube.
3. Rotate tube with seed suspension overhead for 10–15 min.
4. Under the laminar hood, let seeds settle down to the bottom of the tube and remove supernatant by pipetting.
5. Add 1 mL of 100% EtOH to the tube, invert tube 30 times, let seeds settle down to the bottom of the tube and remove supernatant by pipetting. Repeat this step 3–4 times.
6. Incubate seeds in the opened tube located under the laminar hood for ca. 2–3 h until they appear dry. Then close the tube.
7. Transfer six square petri dishes containing LAK medium to laminar hood and let them dry half opened for 1 h.

8. Using a water proof pen, mark four horizontal lines on the dish bottom-side. Lines should be marked along the whole dish axis and 2.5 cm distant above each other. Leave 3 cm space to the lower edge of the dish.
9. Spread dried seeds onto sterile paper.
10. Using a sterile toothpick, slightly wettened at the tip, transfer Arabidopsis ABA biosensor seeds onto LAK medium dishes. Align the seeds along the labeled horizontal lines ca. 4 mm distant from each other.
11. Close petri dishes with micropore tape, wrap them in aluminum foil, and incubate them for 3–7 days at 4 °C for stratification (*see* **Note 6**).
12. Remove the aluminum foil and grow the plant material in a growth cabinet for 4 days. Mount the square petri dishes vertically using a petri dish holder.

3.2 Sample and Imaging Preparations

1. Two days before the microscopy experiment, warm up the bottle with microscope dish medium in a microwave. Under sterile conditions, transfer ca. 7.5 mL of microscope dish medium to a 50 mL tube. From there pipette 200 μL of medium into microscope dish glass-bottom wells of 32 dishes. Make sure that each glass-bottom well is entirely covered with medium. Close the dishes with their upper lid and incubate them for 1 h at RT. After polymerization of the medium, carefully transfer the microscope dishes into a storage box and incubate them overnight at 4 °C (*see* **Note 7**).
2. One day (late afternoon) before the microscopy experiment, incubate the closed microscope dishes for 30–60 min at RT on the lab bench for acclimation. Do not stack the dishes.
3. Check proper expression of the 4-day-old ABA biosensor seedlings at the fluorescence stereo microscope and mark nonfluorescent seedlings (*see* **Note 4**).
4. For each microscope dish, carefully transfer two fluorescent seedlings onto polymerized microscope dish media within the glass-bottom well using fine forceps prior cleaned with 70% EtOH. Make sure that both seedlings are properly aligned and 5 mm distant from each other, with their appearance being as linear as possible (*see* **Note 8**).
5. Using 2–3 cm of micropore tape, close the microscope dishes on the top side to avoid unwanted opening during subsequent incubation.
6. Incubate the microscope dishes in the growth cabinet overnight. Mount the dishes vertically using an appropriate holder.
7. Before the microscopy experiments, prepare an aliquot of plant imaging buffer in a 50 mL tube. Prepare a 1.5 mL tube

containing 1 mL of a 1 mM ABA solution. For this, dilute 10 μL of the 100 mM ABA stock solution into 990 μL of plant imaging buffer.

8. Transfer four microscope dishes with ABA biosensor seedlings to the microscope room. Remove the micropore tape from the dishes and place the dishes horizontally on the table. Immerse the seedlings by carefully pipetting (dropwise) 90 μL of plant imaging buffer to the microscope dishes. Using the same yellow pipet tip, carefully attach the seedlings back to their agarose bed. Make sure that the seedlings are aligned, with their roots as linear as possible. Most importantly, the root tips, including regions 5 mm above, should be as planar as possible so that they appear in the same focus area under the microscope.
9. Incubate the prepared seedlings horizontally for 45–60 min under a desk lamp at ~150 μmol/m^2/s photon fluence rate for recovery.

3.3 Microscopy

1. Turn on the microscope components and load the LAS X software.
2. In the LAS X Configuration tab: Launch the Argon Laser and set the laser intensity to 30%. Allow a warm-up for 45–60 min. Set the resolution to 8-bit.
3. In the LAS X Acquire tab: select Acquisition Mode XYZt, Format 1024 × 178, Speed 400 Hz, Bidirectional X ON, Zoom Factor 0.75, Line Average 2, Frame Average 1, Frame Accumulation 1, Rotation 0 (adjustable), Pinhole 5 AU, Z-Stack (Z-Position 0, Nr. of steps 1), *t* (Time Interval 6 s, Cycle 641 [adjustable]), Visible Laser ON (458 nm at 20% [adjustable]), Objective 10×, PMT1 ON (mTurquoise (mT) emission 465–505 nm, Gain 800 V, Offset −0.4%), PMT3 ON (cpVenus173 (cpV) emission 520–550 nm, Gain 800 V, Offset −0.4%), PMT Trans ON (Gain 200 V [adjustable]) (*see* **Note 9**).
4. Carefully mount a microscope dish with samples onto the microscope stand. In the Live mode, move the microscope stage to place the seedling into the field of view so that the very root tip appears on the right side of the image. Rotate the microscope dish by hand for course radial adjustments and use the Rotation tool for fine adjustments. Adjust the focus and capture an image. Optimize the image acquisition settings for the first experimental replicate and save the settings. Keep these acquisition settings stable for the entire experiment, including all experimental replicates. From the four prepared microscope dishes, select a seedling, of which root fits best into the field of view and which lies completely in the focus area. Make sure, that during the experiment, the root will not grow out of the field of view. Capture an image.

5. In the LAS X Quantify tab: select Intensity Stack Profile, Sort Channels and draw four rectangular regions of interest (ROIs) along the root axis, with ROIs 1 and 2 placed over the root differentiation zone, ROI 3 placed over the elongation zone and ROI 4 placed over the meristematic zone and the root cap. Go to Select a Calculator and choose "Calcium Imaging". In the Analysis tab, select Ratio = Channel.002 (cpV)/Channel.001 (mT). Select Draw Background ROI and choose a rectangular region outside of the root. In the Graphs tab, move the mouse over the graphs, click the right mouse-button and select "Normalize" (*see* **Note 10**).
6. Start the experiment and monitor the cycle numbers.
7. Prepare a 1.5 mL tube with 100 μL of 100 μM ABA, by diluting 10 μL of 1 mM ABA solution (from Subheading 3.2, **step** 7) into 90 μL of plant imaging buffer. Mix well and take up 10 μL of the 100 μM ABA solution into a new yellow pipet tip using a 20 μL pipet.
8. At cycle 40, carefully drop 10 μL of the 100 μM ABA solution to the left- or right side of the imaged root, close to the imaged region. Do not pipet the solution to the side where the second seedling is located. Avoid placing the pipet tip into the light path and avoid touching the microscope dish (*see* **Note 11**).
9. During the running experiment, prepare the next four microscope sample dishes (*see* Subheading 3.2, **steps 8** and **9**).
10. After the experiment is finished, save the acquired data. Indicate in the file name the experiment number, the biosensor name, the treatment time-point and the treatment solution (e.g., exp1_ABAleonSD1-3L21_4min_10uM_ABA.lif).
11. Acquire eight experimental replicates.
12. After all experiments have been completed, wash the microscope dishes. Remove residual medium from the dishes by rinsing them with 70% EtOH. Collect the dishes in a large beaker, top them with water and add a small spoon of SDS. Carefully mix the suspension and incubate it overnight at RT. Rinse the dishes 5× with deionized water and dry them using delicate task wipers. Until the next use, store the dishes in a clean box.

3.4 Image Processing and Analysis

1. Open the LAS X .lif file of the first experiment in Fiji.
2. Select view stack with "Hyperstack", open files individually, color mode "Colorized", split channels and press "OK".
3. Select only the time-lapse series image-stack and press "OK".
4. Close the bright filed channel ($C = 2$).

5. Apply **steps 6–15** to both, ($C = 0$) "mT" and ($C = 1$) "cpV" image stacks.
6. Draw a rectangular ROI in a background region. Press "t" to open the ROI Manager and copy the ROI.
7. In the ROI Manager go to More/Multi Measure, select "measure all slices", "one row per slice" and press "OK". In the Results window, press "Results/Summarize" and memorize the maximum background mean gray value. Unclick the ROI from the image-stack window.
8. Go to Process/Math/Subtract and use as Value the up-rounded maximum background mean gray value + 2. Press "OK". Process all images within the stack (*see* **Note 12**).
9. Go to Process/Filters/Gaussian Blur (Sigma Radius = 1), press "OK" and process all images.
10. Go to Process/Filters/Median (Radius = 1), press "OK" and process all images.
11. Go to Image/Type/32-bit.
12. Press "T" to open Threshold. Use the parameters Default, Red, Dark Background and Stack histogram (*see* **Note 13**). Click "Set" to set the upper threshold level to maximum level (for 8-bit images it is 255) and the lower threshold level so that only the specimen is labeled red (should be close but slightly higher than the background subtraction value). Move with the cursor through the image stack to verify that the selected threshold level is appropriate for all stack images. Press "Apply", select "Set to NaN" and process all images.
13. Go to LUT and select cyan color for the "mT" channel and yellow color for the "cpV" channel.
14. Rotate the image stacks, so that the root tips are aligned toward the gravity vector. For this, go to Image/Transform and select the appropriate parameter (e.g., "Rotate 90 Degrees Right").
15. Go to File/Save As/Tiff and save the "mT" channel as "mT.tif" and the "cpV" channel as "cpV.tif".
16. For the ratio image calculation, go to Process/Image Calculator and select Image1 "cpV", operation "Divide" and Image2 "mT", Select "create new window" and "32-bit (float) result", press "OK" and process all images.
17. Go to LUT and select "royal" color.
18. Go to Image/Adjust/Brightness/Contrast and select "Auto".
19. Go to File/Save As/Tiff and save as "exp1_ABAleonSD1-3L21.tif".
20. Close all windows, including the "ROI Manager" and "Threshold" windows, move the newly generated .tif files

and the original .lif file to a new folder "exp1" and repeat **steps 1–19** for all experimental replicates. Name the latter files and folders respectively (exp2, exp3, . . .).

21. Open the file "exp1_ABAleonSD1-3L21.tif" in Fiji.
22. Go to Plugins/GEFI Analyzer/Vertical Profile/Vertical Profile (raw w/ tip) (*see* **Note 14**). Press "Ctrl + s" to save the file.
23. Go to Plugins/GEFI Analyzer/Vertical Profile/Normalize Vertical Profile. Press "Ctrl + s" to save the file. Close all windows including the ROI Manager.
24. Repeat **steps 21–23** for all experimental replicates.
25. Select all files named "exp(*n*)_ABAleonSD1-3L21_vpr_tip.tif" and open all of them in Fiji. Go to Image/Stacks/Images to Stack, set the Name to "ABAleonSD1-3L21_vpr_tip_stack" and press "OK". Move the cursor through the stack and check whether the experiment order is correct. Press "Ctrl + s" to save the file. Close all windows.
26. Select all files named "exp(*n*)_ABAleonSD1-3L21_vpr_tip_-normalized.tif" and open all of them in Fiji. Go to Image/Stacks/Images to Stack, set the Name to "ABAleonSD1-3L21_vpr_tip_normalized_stack" and press "OK". Move the cursor through the stack and check whether the experiment order is correct. Press "Ctrl + s" to save the file. Close all windows.
27. Keep the Vertical Profile stack files and delete the single Vertical Profile images.
28. Open the file "ABAleonSD1-3L21_vpr_tip_stack.tif" in Fiji. Go to Image/Stacks/Z Project, select Start slice "1", Stop slice "8", Projection type "Average Intensity", and press "OK".
29. Go to LUT and select "royal" color.
30. Go to Image/Adjust/Brightness/Contrast, select "Auto" and note the minimum and maximum values.
31. Press "Ctrl + s" to save the file (*see* **Note 15** for extracting the LUT calibration bar).
32. Go to Image/Type/RGB color. Go to File/Save As/Tiff and save as "AVG_ABAleonSD1-3L21_vpr_tip_stack_rgb.tif". This image will be included in Fig. 2a.
33. Repeat **steps 28–32** using the file "ABAleonSD1-3L21_vpr_tip_normalized_stack.tif", and name subsequently generated files accordingly. The image "AVG_ABAleonSD1-3L21_vpr_tip_stack_normalized_rgb.tif" will be included in Fig. 2b.
34. Close all open images and windows in Fiji.

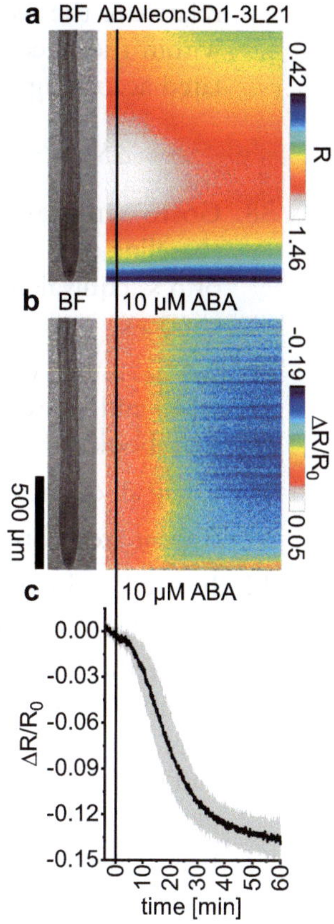

Fig. 2 ABAleonSD1-3L21 analyses in 5-day-old Arabidopsis roots in response to 10 μM ABA. (**a**) Average vertical response profile of the ABAleonSD1-3L21 emission ratio (R = cpV/mT) in response to 10 μM ABA ($n = 8$). (**b**) Average normalized vertical response profile of the ABAleonSD1-3L21 emission ratio change ($\Delta R/R_0 = [R(t) - R_{AVG(baseline)}]/R_{AVG(baseline)}$) in response to 10 μM ABA ($n = 8$). (**c**) Whole root image ABAleonSD1-3L21 emission ratio change in response to 10 μM ABA (AVG ± SD, $n = 8$). Image data were acquired for a duration of 64 min at a time interval of 6 s and 10 μM ABA was applied at $t = 0$ min

35. Open Microsoft Excel and generate a new blank workbook. Add the following headers to the columns A–K: time [min], exp1, exp2, ..., exp8, AVG, SD. In the "time" column, set the time scale from −4.0, −3.9, −3.8, ..., up to 60 min. These values depend on the imaging Time Interval of 6 s (0.1 min), the length of the baseline recording (4 min) and the length of the treatment recording (60 min).
36. Open the file "exp1_ABAleonSD1-3L21.tif" in Fiji.
37. Go to Plugins/GEFI Analyzer/DEAD Plot (baseline slope ±5). Besides other parameters (*see* GEFI Analyzer README), the DEAD Plot calculates the time-dependent signal change

relative to the average signal (S_{AVG}) of the selected baseline and within the selected ROI. The signal change values are calculated according to the formula:

$$\Delta S/S_0 = \left[S(t) - S_{AVG(baseline)}\right]/S_{AVG(baseline)}.$$

38. Go to setup and make sure that "X start" is set to "−4.0" (baseline of 4 min) and "X step" is set to "0.1" (Time Interval of 6 s, equivalent to 0.1 min) (*see* **Note 16**). Close the setup window.
39. Go to results, copy the "mean"-column data from the Results table and paste them into the "exp1"-column of the Excel workbook.
40. Close all opened images and windows in Fiji.
41. Repeat **steps 36–40** for all eight experiments.
42. In the Excel workbook, calculate the average (AVG) and standard deviation (SD) of all experiments and for each time-point.
43. From the Excel workbook, import the time, AVG and SD columns into OriginLab—OriginPro or another graphing software and plot the data into a graph as shown in Fig. 2c. Export and save the graph in a vector graphic-based format, for example, "ABAleonSD1-3L21.emf".
44. From the eight experiments, choose one representative experiment and open the respective "exp*(n)*_ABAleonSD1-3L21.tif" file in Fiji. Make sure that the ROI Manager is closed, and go to Plugins/GEFI Analyzer/Vertical Profile/Vertical Profile (raw w/ tip). This command reopens the ROI Manager and generates ROIs based on the Plugin/GEFI Analyzer/Tip Detection Settings (*see* **Note 17**). Close all open images, but keep the ROI Manager open.
45. Open the respective LAS X "exp*(n)*_ABAleonSD1-3L21_4min_10uM_ABA.lif" file in Fiji.
46. Select view stack with "Hyperstack", open files individually, color mode "Colorized", split channels and press "OK".
47. Select only the time-lapse series image-stack and press "OK".
48. From the newly opened image stacks close all, except of the bright filed channel ($C = 2$).
49. Rotate the bright field image stack, so that the root tip is aligned toward the gravity vector. Use the same parameters as in **step 14**.
50. Go to Image/Stacks/Tools/Slice Keeper and select First Slice 41, Last Slice 41, Increment 1. This command will extract the image-stack frame (Slice) at $t = 0$ min (*see* **Note 18**).

51. For the newly opened image, select the ROI "0041-..." in the ROI Manager and go to Image/Crop.
52. Go to Image/Adjust/Brightness/Contrast and adjust the parameters to taste.
53. Go to Image/Type/RGB Color and save the image as "exp*(n)*_BF". This image will be included in Fig. 2a, b.
54. Close Fiji.
55. Import images, graph and calibration bar generated in **steps 32, 33, 43** and **53** (*see* **Note 15**) into Adobe Illustrator, or another graphic design tool and assemble a figure as shown in Fig. 2. For the calibration bar use the values noted in **step 30**.

4 Notes

1. The choice of the right ABA biosensor line depends on (1) the experimental question, and (2) the tissue of interest. Compared to ABACUS1-2μ, ABAleons exhibit a higher ABA affinity but a lower dynamic range (Table 1). Therefore, ABAleons likely exhibit a better response in tissues with low basal [ABA]. ABACUS1-2μ might respond better in tissues with high basal [ABA], for example in guard cells. The original transgenic lines expressing ABAleon2.1, ABAleon2.15, and ABACUS1-2μ [5, 6] exhibit either noticeable silencing issues or low expression levels, respectively. These drawbacks have been mainly overcome in more recently generated transgenic Arabidopsis lines expressing ABAleon2.15, ABAleonSD1-3L21, or ABACUS1-2μ [7]. Transgenic Arabidopsis lines expressing ABAleon2.15 under control of the guard cell preferential promoter *pGC1* [19] are also recommended for guard cell ABA analyses. To avoid seed harvesting of potentially silenced plants, it is strongly recommended to check the fluorescence of ABA biosensor plants already at an early seedling stage before transfer to soil, and again when the plants are 3 weeks old. Also, seeds should be harvested in single-plant batches, and each seed batch should be tested for proper seed germination and biosensor expression.
2. Before taking aliquots of LAK macros, micros and FeNaEDTA, carefully mix the stock solutions to avoid unwanted precipitates. As an alternative to the LAK medium, half-strength Murashige and Skoog medium including vitamins could be used.
3. Microscope dishes can be reused after washing.
4. The use of a stereo microscope is optional. However, dependent on the percentage of silenced ABA biosensor seedlings, selection of fluorescing seedlings prior to imaging helps to avoid further processing of nonfluorescent seedlings.

5. This protocol describes FRET-based ABA biosensor imaging using an inverted Leica SP8 confocal laser scanning microscope. However, any inverted microscope that allows for mT excitation and rapid mT and cpV emission analyses would be applicable for similar analyses. If the microscope sample holder is not suitable for mounting glass-bottom dishes, use a 1 mm thick aluminum disc on which the glass-bottom dishes can be fixed with double-sided tape. The aluminum disc should have an external diameter to fit onto the microscope sample holder and an inner hole-diameter that leaves the glass-bottom well free for microscopy access.
6. Because ABACUS1-2µ expressing plants are ABA hypersensitive [5] and seeds are more dormant, they require 7 days for stratification.
7. Residual microscope dish medium stored in a 100 mL bottle can be reused until emptied. To ensure that a suitable seedling can be found for each experimental replicate, it is recommended to prepare four microscope dishes per experiment. Thirty-two dishes should last for eight experimental replicates.
8. For the seedling-transfer to microscope dishes, select each seedling based on its size (should fit into the glass-bottom well) and the shape of its primary root (should be as linear as possible). Remove the selected seedling from the LAK medium by carefully lifting it at the cotyledons using fine forceps. Close the LAK medium dish to avoid dehydration of the other seedlings. Proceed with laying the seedling onto the microscope dish medium with the root tip first. Slightly pull the seedling upward to enable a linear positioning. As the seedling continues to grow until the next day, make sure that the root will not grow out of the glass-bottom well. Transfer a second seedling to the microscope dish 5 mm distant but parallel to the first seedling. Immediately close the microscope dish. After the seedling transfer has been completed, dispose of the remaining plant material by autoclaving.
9. The imaging settings depend on the expression level of the biosensor and the quality of the microscope components. They need to be carefully adjusted prior to the time-lapse imaging. It is recommended to set the laser power as low as possible and not higher than 40%. Gain settings around 800 V are generally recommended. Higher gain settings may result in increased background noise. Acquire test images of the tissue of interest using the proposed settings and check whether the detectors are overexposed. If yes, lower the laser power before reducing the detector gain. Keep in mind that fluorescence emission intensities change during the time-lapse experiment. Make sure that the detectors will not be overexposed during

the entire experiment. If the acquired signal intensity is very low, increase the laser power (maximum 40%) and gain (maximum 950 V). Alternatively set Line Average 1 and Frame Accumulation 2. Adjusting the offset is great for eliminating background noise below a certain threshold. Make sure that the biosensor fluorescence signal is not affected by the imposed offset settings. Optimized offset settings greatly ease subsequent image processing procedures such as background subtraction and thresholding. After the imaging settings have been optimized, they must be kept identical for all experimental replicates. The duration of the experiment can be extended up to 84 min. After this time the imaged sample may have dried out.

10. The LAS X Quantify tool is convenient for live monitoring of the running experiment. It is not intended to be used for image analyses. Alternative to the "Calcium Imaging" tool, select the entire image as an ROI and Select a Calculator "ROI/Channel". Go to the Analysis tab, and select Graph = Channel.002/ROI.01/Channel.001/ROI.01.
11. Through the application of 10 μM ABA at cycle 40 and a total number of 641 cycles at a time interval of 6 s, the experiment is divided into a baseline-recording period of 4 min and an ABA response-recording period of 60 min. Although these settings can be adjusted to taste, it is strongly recommended to keep these parameters stable for all experiments. Such consistency will ease later data analysis procedures.
12. If there is still a ROI selected in the image, the background subtraction will be applied only to the ROI. It is highly recommended to use identical background subtraction settings for each experimental replicate and the respective fluorescence channel.
13. In order that the Stack histogram command works, it is required to unclick and reclick it. It is highly recommended to use identical threshold settings for each experimental replicate and the respective fluorescence channel.
14. Make sure that the ROI Manager window has been closed beforehand. This command detects the root tip and writes for each image-stack frame a ROI into the ROI manager that encompasses a region from the root tip up to a height which is defined in the Plugin/GEFI Analyzer/Tip Detection Settings. For the imposed imaging settings, the recommended parameters are: Image Height 896 px, Not NaN Pixel Count 20 and Tip Offset 5.
15. For generating the royal LUT calibration bar, go to Analyze/Tools/Calibration Bar and adjust the settings to taste. Increase the Zoom factor to 6, or as much as possible, and press "OK".

Go to Image/Type/RGB Color. Using a rectangular ROI, select only the color bar and go to Image/Crop. Save the cropped color bar as "royal_LUT_calibration_bar.tif".

16. "X start", "X step" and the time scale values in the Excel workbook need to be set according to the image acquisition time-lapse parameters and the time-point of the treatment (set as $t = 0$ min).
17. Use the same tip detection settings as before.
18. It is also possible to extract other slices. The $t = 0$ min slice number depends on the microscope time-lapse imaging settings.

Acknowledgments

I thank Prof. Hannes Kollist and Dr. Ebe Merilo (University of Tartu, Institute of Technology) for supporting me and giving me the opportunity to continue my research. This work was supported by Estonian Research Council grants PRG433 to HK and PRG719 to EM.

References

1. Allen GJ, Kwak JM, Chu SP, Llopis J, Tsien RY, Harper JF, Schroeder JI (1999) Cameleon calcium indicator reports cytoplasmic calcium dynamics in Arabidopsis guard cells. Plant J 19(6):735–747
2. Uslu VV, Grossmann G (2016) The biosensor toolbox for plant developmental biology. Curr Opin Plant Biol 29:138–147
3. Hilleary R, Choi WG, Kim SH, Lim SD, Gilroy S (2018) Sense and sensibility: the use of fluorescent protein-based genetically encoded biosensors in plants. Curr Opin Plant Biol 46: 32–38
4. Walia A, Waadt R, Jones AM (2018) Genetically encoded biosensors in plants: pathways to discovery. Annu Rev Plant Biol 69:497–524
5. Jones AM, Danielson JA, Manojkumar SN, Lanquar V, Grossmann G, Frommer WB (2014) Abscisic acid dynamics in roots detected with genetically encoded FRET sensors. elife 3:e01741
6. Waadt R, Hitomi K, Nishimura N, Hitomi C, Adams SR, Getzoff ED, Schroeder JI (2014) FRET-based reporters for the direct visualization of abscisic acid concentration changes and distribution in Arabidopsis. elife 3:e01739
7. Waadt R, Köster P, Andrés Z, Waadt C, Bradamante G, Lampou K, Kudla J, Schumacher K (2020) Dual-reporting transcriptionally linked genetically encoded fluorescent indicators resolve the spatiotemporal coordination of cytosolic abscisic acid and second messenger dynamics in Arabidopsis. Plant Cell 32(8):2582–2601
8. Ma Y, Szostkiewicz I, Korte A, Moes D, Yang Y, Christmann A, Grill E (2009) Regulators of PP2C phosphatase activity function as abscisic acid sensors. Science 324(5930): 1064–1068
9. Park SY, Fung P, Nishimura N, Jensen DR, Fujii H, Zhao Y, Lumba S, Santiago J, Rodrigues A, Chow TF, Alfred SE, Bonetta D, Finkelstein R, Provart NJ, Desveaux D, Rodriguez PL, McCourt P, Zhu JK, Schroeder JI, Volkman BF, Cutler SR (2009) Abscisic acid inhibits type 2C protein phosphatases via the PYR/PYL family of START proteins. Science 324(5930): 1068–1071
10. Melcher K, Ng LM, Zhou XE, Soon FF, Xu Y, Suino-Powell KM, Park SY, Weiner JJ, Fujii H, Chinnusamy V, Kovach A, Li J, Wang Y, Li J, Peterson FC, Jensen DR, Yong EL, Volkman BF, Cutler SR, Zhu JK, Xu E (2009) A gate-latch-lock mechanism for hormone signalling by abscisic acid receptors. Nature 462(7273): 602–608
11. Miyazono K, Miyakawa T, Sawano Y, Kubota K, Kang HJ, Asano A, Miyauchi Y, Takahashi M, Zhi Y, Fujita Y, Yoshida T,

Kodaira KS, Yamaguchi-Shinozaki K, Tanokura M (2009) Structural basis of abscisic acid signalling. Nature 462(7273):609–614
12. Yin P, Fan H, Hao Q, Yuan X, Wu D, Pang Y, Yan C, Li W, Wang J, Yan N (2009) Structural insights into the mechanisms of abscisic acid signaling by PYL proteins. Nat Struct Mol Biol 16(12):1230–1236
13. Hochreiter B, Garcia AP, Schmid JA (2015) Fluorescent proteins as genetically encoded FRET biosensors in life sciences. Sensors 15(10):26281–26314
14. Bajar BT, Wang ES, Zhang S, Lin MZ, Chu J (2016) A guide to fluorescent protein FRET pairs. Sensors 16(9):1488
15. Waadt R, Hsu PK, Schroeder JI (2015) Abscisic acid and other plant hormones: methods to visualize distribution and signaling. BioEssays 37(12):1338–1349
16. Dupeux F, Santiago J, Betz K, Twycross J, Park SY, Rodriguez L, Gonzalez-Guzman M, Jensen MR, Krasnogor N, Blackledge M, Holdsworth M, Cutler SR, Rodriguez PL, Márquez JA (2011) A thermodynamic switch modulates abscisic acid receptor sensitivity. EMBO J 30(20):4171–4184
17. Nishimura N, Hitomi K, Arvai AS, Rambo RP, Hitomi C, Cutler SR, Schroeder JI, Getzoff ED (2009) Structural mechanism of abscisic acid binding and signaling by dimeric PYR1. Science 326(5958):1373–1379
18. Batool S, Uslu VV, Rajab H, Ahmad N, Waadt R, Geiger D, Malagoli M, Xiang CB, Hedrich R, Rennenberg H, Herschbach C, Hell R, Wirtz M (2018) Sulfate is incorporated into cysteine to trigger ABA production and stomatal closure. Plant Cell 30(12): 2973–2987
19. Hsu PK, Takahashi Y, Munemasa S, Merilo E, Laanemets K, Waadt R, Pater D, Kollist H, Schroeder JI (2018) Abscisic acid-independent stomatal CO_2 signal transduction pathway and convergence of CO_2 and ABA signaling downstream of OST1 kinase. Proc Natl Acad Sci U S A 115(42):E9971–E9980
20. Zhang L, Takahashi Y, Hsu PK, Kollist H, Merilo E, Krysan PJ, Schroeder JI (2020) FRET kinase sensor development reveals SnRK2/OST1 activation by ABA but not by MeJA and high CO_2. elife 9:e56351
21. Zhao Y, Araki S, Wu J, Teramoto T, Chang YF, Nakano M, Abdelfattah AS, Fujiwara M, Ishihara T, Nagai T, Campbell RE (2011) An expanded palette of genetically encoded Ca^{2+} indicators. Science 333(6051):1888–1891
22. Shen Y, Rosendale M, Campbell RE, Perrais D (2014) pHuji, a pH-sensitive red fluorescent protein for imaging of exo- and endocytosis. J Cell Biol 207(3):419–432
23. Waadt R, Krebs M, Kudla J, Schumacher K (2017) Multiparameter imaging of calcium and abscisic acid and high-resolution quantitative calcium measurements using R-GECO1-mTurquoise in Arabidopsis. New Phytol 216(1):303–320
24. Christmann A, Weiler EW, Steudle E, Grill E (2007) A hydraulic signal in root-to-shoot signalling of water shortage. Plant J 52(1): 167–174
25. Ikegami K, Okamoto M, Seo M, Koshiba T (2009) Activation of abscisic acid biosynthesis in the leaves of *Arabidopsis thaliana* in response to water deficit. J Plant Res 122(2): 235–243
26. Takahashi F, Suzuki T, Osakabe Y, Betsuyaku S, Kondo Y, Dohmae N, Fukuda H, Yamaguchi-Shinozaki K, Shinozaki K (2018) A small peptide modulates stomatal control via abscisic acid in long-distance signalling. Nature 556(7700):235–238
27. Endo A, Sawada Y, Takahashi H, Okamoto M, Ikegami K, Koiwai H, Seo M, Toyomasu T, Mitsuhashi W, Shinozaki K, Nakazono M, Kamiya Y, Koshiba T, Nambara E (2008) Drought induction of Arabidopsis 9-cis-epoxycarotenoid dioxygenase occurs in vascular parenchyma cells. Plant Physiol 147(4): 1984–1993
28. Bauer H, Ache P, Lautner S, Fromm J, Hartung W, Al-Rasheid KA, Sonnewald S, Sonnewald U, Kneitz S, Lachmann N, Mendel RR, Bittner F, Hetherington AM, Hedrich R (2013) The stomatal response to reduced relative humidity requires guard cell-autonomous ABA synthesis. Curr Biol 23(1):53–57
29. Kuromori T, Seo M, Shinozaki K (2018) ABA transport and plant water stress responses. Trends Plant Sci 23(6):513–522
30. Kang J, Yim S, Choi H, Kim A, Lee KP, Lopez-Molina L, Martinoia E, Lee Y (2015) Abscisic acid transporters cooperate to control seed germination. Nat Commun 6:8113
31. Behera S, Kudla J (2013) High-resolution imaging of cytoplasmic Ca^{2+} dynamics in Arabidopsis roots. Cold Spring Harb Protoc 7: 670–674
32. Krebs M, Schumacher K (2013) Live cell imaging of cytoplasmic and nuclear Ca^{2+} dynamics in Arabidopsis roots. Cold Spring Harb Protoc 8:776–780
33. Swanson SJ, Gilroy S (2013) Imaging changes in cytoplasmic calcium using the Yellow Cameleon 3.6 biosensor and confocal microscopy. Methods Mol Biol 1009:291–302

34. Okumoto S (2014) Quantitative imaging approaches for small-molecule measurements using FRET sensors in plants. Methods Mol Biol 1083:55–64

35. Rizza A, Walia A, Tang B, Jones AM (2019) Visualizing cellular gibberellin levels using the nlsGPS1 Förster resonance energy transfer (FRET) biosensor. J Vis Exp 143:e58739

36. Barragán V, Leidi EO, Andrés Z, Rubio L, De Luca A, Fernández JA, Cubero B, Pardo JM (2012) Ion exchangers NHX1 and NHX2 mediate active potassium uptake into vacuoles to regulate cell turgor and stomatal function in Arabidopsis. Plant Cell 24(3):1127–1142

37. Allen GJ, Chu SP, Harrington CL, Schumacher K, Hoffmann T, Tang YY, Grill E, Schroeder JI (2001) A defined range of guard cell calcium oscillation parameters encodes stomatal movements. Nature 411(6841):1053–1057

38. Schindelin J, Arganda-Carreras I, Frise E, Kaynig V, Longair M, Pietzsch T, Preibisch S, Rueden C, Saalfeld S, Schmid B, Tinevez JY, White DJ, Hartenstein V, Eliceiri K, Tomancak P, Cardona A (2012) Fiji: an open-source platform for biological-image analysis. Nat Methods 9(7):676–682

Chapter 12

In Situ Observation of Abscisic Acid Distribution in Major Crop Species by Immunofluorescence Labeling

Mo-Xian Chen, Chong-Chong Lu, Jianhua Zhang, and Ying-Gao Liu

Abstract

Immunofluorescent approach uses fluorophore-conjugated antibodies to target molecules of interest for the observation of their distribution and microenvironment at cellular or tissue level. In connection with modern fluorescence microscopy, immunofluorescence has high sensitivity and specificity, resulting in a wide range of applications in animal studies. However, the protocols of immunofluorescence are seldom reported. To this end, we describe an optimized protocol for the detection of plant hormone abscisic acid (ABA) in monocot crop species, including maize and rice. By using immunofluorescence technique provided from this protocol, investigation of ABA distribution will continue to deeper our insights in crop developments and stress responses.

Key words Abscisic acid, Fluorescence microscopy, Immunofluorescence, *Oryza sativa*, Tissue imaging, *Zea mays*

1 Introduction

Immunofluorescence (IF) is an immunochemical technique that permits visualization of proteins or small molecules at tissue or cellular level [1]. It has been widely used in both animal and plant studies [2–5]. IF could be applied to different sample types, including cell suspension, cultured cells, tissue sections, and entire organism [6]. The whole technique is initiated by fixation, an important step prior to IF staining to immobilize antigens of interest and preserve cellular morphology at the same time [6]. Thus, choice of fixatives and optimization of steps for fixation are crucial for each target antigen. Common cross-linking reagents, such as formaldehyde and glutaraldehyde are used to form methylene cross-links within cells [6, 7]. While organic solvents are often used to remove excessive water and lipids, aiming to denature cellular structure for

Takuya Yoshida (ed.), *Abscisic Acid: Methods and Protocols*,
Methods in Molecular Biology, vol. 2462, https://doi.org/10.1007/978-1-0716-2156-1_12,

further experimental steps. Up to date, two IF approaches can be selected, naming as direct/primary and indirect/secondary IF [6, 7]. The difference between these two approaches is resulted from the attachment of labeled fluorophore to primary (direct IF) or secondary (indirect IF) antibodies [6]. The subsequent detection of fluorescent signal is based on the development of fluorescence microscopy [7]. The fluorescent signal can be observed by either a conventional epifluorescence or a confocal microscope.

As the most widely used imaging analysis to detect subcellular localization of target molecules, the advantages and limitation of IF has been extensively discussed [6]. Nowadays, multiple choices of fluorescent moieties with different excitation and emission wavelengths are available commercially, facilitating the simultaneous detection of multiple targets by using combinations of specific antibodies. Furthermore, IF is able to detect multiple targets of interest through repeated destaining and restaining cycles [8]. However, to achieve this goal, the selection of proper antibodies and emission spectra of fluorophores are critical. Furthermore, common fluorescence problems such as photobleaching and autofluorescence will interfere the results. In addition, although slide preparation of IF may encounter difficulties such as permeabilization, during fixation procedures, IF uses antibodies to achieve its high specificity and provides an alternative way other than ectopic expression using plasmids.

Abscisic acid (ABA) is one of the classic plant hormones initially characterized to play crucial role in response to water deprivation. ABA is not only a stress-responsive hormone, but also involved in development under nonstress conditions [9, 10] mediating molecular regulations at both transcriptional and post-transcriptional levels [11–13]. Thus, the investigation of in situ distribution of ABA will facilitate the understanding of its molecular mechanism during plant developmental processes and stress responses. The IF approach to detect ABA has been applied to model dicot plant, *Arabidopsis thaliana* [14]. Recently, we have applied this technique in maize tissues to reveal possible mechanism of ABA in lateral root development and seed germination [15, 16]. In this chapter, we summarized the protocol used to observe ABA distribution in monocot crop species, such as maize and rice. The step-by-step protocol provided here is optimized for detection ABA molecules in several tissue types of these two plant species. The corresponding experimental procedures including plant growth condition, tissue preparation, fixation, the use of fluorophore-conjugated antibodies, and precautions are listed in the following sections.

2 Materials

2.1 Chemicals and Solutions

1. Sterile distilled water.
2. 10 mM sodium phosphate-buffered saline (PBS, pH 7.2).
3. Anti-ABA polyclonal antibody.
4. Goat anti-rabbit IgG secondary antibody conjugated to Alexa Fluor 488 (*see* **Note 1**).
5. Citifluor AF1.
6. Nonfat dried milk.
7. 75% ethanol.
8. 5% sodium hypochlorite.
9. 1/2-strength Hoagland's nutrient solution.
10. Stationary liquid (SL) solution: 10 mM PBS (pH 7.2), 0.2% (v/v) glutaraldehyde, 4% (w/v) paraformaldehyde, 2% (w/v) 1-ethyl-3-(3-dimethylaminopropyl) carbodiimide hydrochloride.
11. Buffer A: 10 mM PBS (pH 7.2), 0.2% (w/v) cellulase, 0.2% (w/v) pectinase, 3% (w/v) nonfat dried milk, 0.1% Triton X-100.
12. Buffer B: 10 mM PBS, 3% (w/v) nonfat dried milk, 0.05% Tween 20.
13. Mounting solution: 10 mM PBS (pH 7.2).

2.2 Apparatus, Equipment, and Analytical Software

1. Shaker.
2. Scalpel blades.
3. Scissors.
4. Confocal laser-scanning microscope.
5. 24-well incubation plate.
6. Vacuum infiltrator.
7. Fume hood.
8. Round petri dish with diameter = 10 cm.
9. Sterile, moist absorbent cotton gauze.
10. Plant growth chambers.
11. A4 paper.
12. ImageJ software.

2.3 Plant Materials

All plant materials used for immunolabeling experiments are freshly prepared before the staining protocol.

1. Maize roots and seeds.
2. Rice leaves.

3 Methods

3.1 Plant Materials and Growth Condition

1. Maize plants are grown at optimum condition (26 °C, 70% relative humidity, 12/12 h light/dark). First, all maize seeds are surface sterilized with 75% ethanol for 1 min, and then transferred into 5% sodium hypochlorite for 10 min. After that the seeds are washed by sterile distilled water for three times. And then the seeds are placed in a round petri dish (diameter = 10 cm) and covered with sterile, moist absorbent cotton gauze for germination.
2. After germination for 3 days, the maize seedlings with similar root lengths are transferred to 1/2-strength Hoagland's nutrient solution, for an additional 3–4 days.
3. Rice plants are grown at optimum condition in growth chambers (28 °C, 70% relative humidity, 16/8 h light/dark).

3.2 Tissue Preparation for Immunochemistry

1. Select and harvest appropriate maize primary roots from at least six different plants, cutting them into round transverse pieces with 100–300 μm thick by using two scalpel blades with 1–3 sheets of A4 paper sandwiched between them (*see* **Note 2**).
2. Maize seeds are longitudinal or cross sectioned by using scalpel blades after 2–3 days germination. All the sectioned half-seeds are used for the following experiments.
3. Rice leaves from 3-week-old rice seedlings are dissected into 1 × 1 cm small squares by scissors. Leave veins are subsequently removed and the resulting samples are used for the following experiments (*see* **Note 3**).

3.3 Detection of ABA Fluorescence in Maize and Rice Tissues

1. Put the sectioned plant samples into 24-well incubation plate together with 0.5–1 mL of SL solution (*see* **Note 4**). The samples are vacuum infiltrated at −0.06 to −0.07 MPa for 1–2 h by vacuum infiltrator in the fume hood (*see* **Note 5**) and then shaken at 4 °C overnight.
2. The sectioned plant samples are washed for three times with 10 mM PBS (pH 7.2) and then hyalinized in Buffer A for blocking at least 1 h (*see* **Note 6**).
3. The corresponding tissue samples, including negative controls (i.e., samples subsequently incubated in Buffer B without the addition of antibody) (*see* **Note 7**), are washed for three times again with 10 mM PBS (pH 7.2) and incubated in Buffer B with 1/2000 anti-ABA polyclonal antibodies (*see* **Note 8**) on a shaker at 4 °C overnight (*see* **Note 9**).
4. The samples are washed until they become clear and transferred to a 24-well incubation plate with 0.5–1 mL of Buffer B with goat anti-rabbit IgG secondary antibody conjugated to Alexa

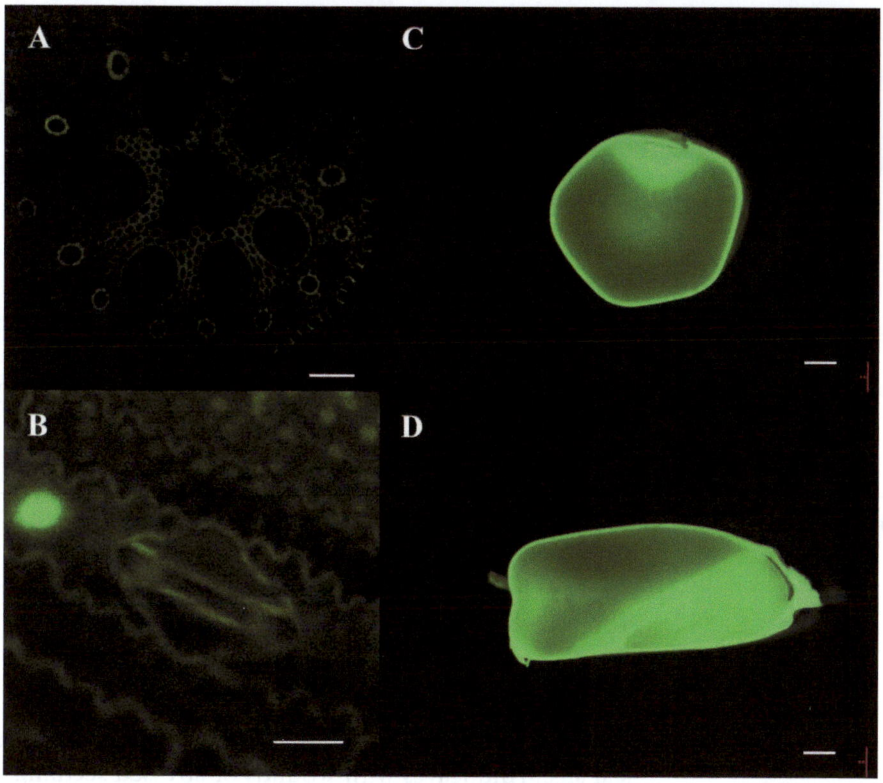

Fig. 1 Sample presentation of ABA distribution in crop tissues detected by immunofluorescence. Immunofluorescent detection of in situ ABA distribution in cross-sectioned maize root (**a**), rice leaf (**b**), cross (**c**), and longitudinal (**d**) sectioned maize seeds. Scale bars = 50 μm (**a**), 5 μm (**b**), 1 mm (**c, d**)

Fluor 488 (one drop of antibody/mL Buffer B) for at least 2 h at room temperature.

5. The samples are washed with 10 mM PBS and blocked with Citifluor AF1.
6. A confocal laser-scanning microscope is ultimately used for microscopic imaging (*see* **Note 10**). Via an argon laser, green fluorescent protein (GFP) is excited at a wavelength of 488 nm and measured at 505–530 nm with a bandpass filter (green) (Fig. 1) (*see* **Note 11**). Short storage of unobserved sample slides in darkness are recommended (*see* **Note 12**).

3.4 Quantification of Fluorescence Intensity

1. The quantification of fluorescence is carried out by using ImageJ software.
2. Open the target image taken by the confocal microscope.
3. Adjust the gray scale to 8 bit and flip the photo in black and white.
4. Choose menu "analyse-calibrate" to adjust the correct photo light density.

5. Choose menu "Uncalibrate OD" and check "global calibration."
6. Select proper unit for the measurement in menu "set scale" and check the "global" option.
7. Select options "area integrated density" and "limit the threshold" below "analyse measure."
8. Choose menu "image-adjust-threshold" and click the measure button.
9. Fluorescence intensity could be measured either transversely or vertically across the sample image. Three measurements are carried out as repeats to calculate the error bars by using Student *t*-test.

4 Notes

1. The excitation and emission wavelengths of Alexa Fluor 488 are 488 nm and 505–530 nm, respectively.
2. At least six maize primary roots are used for each elicitor treatment. Collect at least 3–4 pieces from each plant.
3. Ensure to choose rice seedlings with similar growth stage and leaf size. And leaves with wounds or stressed seedlings are not selected.
4. This is an important step to cross-link the carboxyl group of ABA for in situ fixation.
5. The duration of fixation is based on the permeability of plant tissues, which is always determined by the size and type of specific tissues. For example, rice leaves are recommended to have a 2 h fixation due to its low permeability of its outer wax layer.
6. This step is to prevent nonspecific interactions between antibodies and target tissue.
7. The negative control, that is, tissues incubated in pure Buffer B without the antibody, is required to exclude the nonspecific signals or tissue autofluorescence detected later on.
8. To get the ideal concentration of a particular antibody, serial dilution experiments are suggested to carry out first.
9. Prevention of sample drying is crucial during these steps.
10. To prevent formation of bubbles during the slide preparation, add 300–500 μL of mounting solution on the tissue and gently putting on the cover slip.
11. To prevent autofluorescence, negative controls (*see* **Note 7**) are used to adjust the background fluorescence before examining the samples conjugated with antibodies.

12. Long storage of slides are not recommended. Short storage is suggested to put the unobserved slides in humid condition and darkness to prevent drying and fluorophore fading of samples, respectively.

Acknowledgments

This work was supported by the Science Technology and Innovation Committee of Shenzhen (GJHZ20190821160401654), the National Natural Science Foundation of China (NSFC32001452), the Natural Science Foundation of Jiangsu Province (SBK2020042924), and Hong Kong Research Grant Council (AoE/M-05/12, AoE/M-403/16, GRF CUHK14160516, 14177617, 12100318).

References

1. Mori H, Cardiff RD (2016) Methods of immunohistochemistry and immunofluorescence: converting invisible to visible. Methods Mol Biol 1458:1–12. https://doi.org/10.1007/978-1-4939-3801-8_1
2. Lightbody ED, Nicol CJB (2019) Immunofluorescence labeling of nuclear receptor expression in formalin-fixed, paraffin-embedded tissue. Methods Mol Biol 1966:101–105. https://doi.org/10.1007/978-1-4939-9195-2_8
3. Gulliver LSM (2019) In vivo quantitation of estrogen receptor β subtype expression in ovarian surface epithelium using immunofluorescence profiling and confocal microscopy. Methods Mol Biol 1966:27–38. https://doi.org/10.1007/978-1-4939-9195-2_4
4. Lü B, Chen F, Gong ZH, Xie H, Zhang JH, Liang JS (2007) Intracellular localization of integrin-like protein and its roles in osmotic stress-induced abscisic acid biosynthesis in Zea mays. Protoplasma 232(1–2):35–43. https://doi.org/10.1007/s00709-007-0278-3
5. Qu L, Wu C, Zhang F, Wu Y, Fang C, Jin C, Liu X, Luo J (2016) Rice putative methyltransferase gene OsTSD2 is required for root development involving pectin modification. J Exp Bot 67(18):5349–5362. https://doi.org/10.1093/jxb/erw297
6. Im K, Mareninov S, Diaz MFP, Yong WH (2019) An introduction to performing immunofluorescence staining. Methods Mol Biol 1897:299–311. https://doi.org/10.1007/978-1-4939-8935-5_26
7. Pemberton LF (2014) Preparation of yeast cells for live-cell imaging and indirect immunofluorescence. Methods Mol Biol 1205:79–90. https://doi.org/10.1007/978-1-4939-1363-3_6
8. Stack EC, Wang C, Roman KA, Hoyt CC (2014) Multiplexed immunohistochemistry, imaging, and quantitation: a review, with an assessment of Tyramide signal amplification, multispectral imaging and multiplex analysis. Methods 70(1):46–58. https://doi.org/10.1016/j.ymeth.2014.08.016
9. Yoshida T, Christmann A, Yamaguchi-Shinozaki K, Grill E, Fernie AR (2019) Revisiting the basal role of ABA - roles outside of stress. Trends Plant Sci 24(7):625–635. https://doi.org/10.1016/j.tplants.2019.04.008
10. Yoshida T, Obata T, Feil R, Lunn JE, Fujita Y, Yamaguchi-Shinozaki K, Fernie AR (2019) The role of abscisic acid signaling in maintaining the metabolic balance required for arabidopsis growth under nonstress conditions. Plant Cell 31(1):84–105. https://doi.org/10.1105/tpc.18.00766
11. Chen MX, Mei LC, Wang F, Boyagane Dewayalage IKW, Yang JF, Dai L, Yang GF, Gao B, Cheng CL, Liu YG, Zhang J, Hao GF (2020) PlantSPEAD: a web resource towards comparatively analysing stress-responsive expression of splicing-related proteins in plant. Plant Biotechnol J 19:227. https://doi.org/10.1111/pbi.13486
12. Chen MX, Zhang Y, Fernie AR, Liu YG, Zhu FY (2020) SWATH-MS-based proteomics: strategies and applications in plants. Trends Biotechnol 39:433. https://doi.org/10.1016/j.tibtech.2020.09.002

13. Chen MX, Zhu FY, Gao B, Ma KL, Zhang Y, Fernie AR, Chen X, Dai L, Ye NH, Zhang X, Tian Y, Zhang D, Xiao S, Zhang J, Liu YG (2020) Full-length transcript-based proteogenomics of rice improves its genome and proteome annotation. Plant Physiol 182(3): 1510–1526. https://doi.org/10.1104/pp.19.00430
14. Ondzighi-Assoume CA, Chakraborty S, Harris JM (2016) Environmental nitrate stimulates abscisic acid accumulation in arabidopsis root tips by releasing it from inactive stores. Plant Cell 28(3):729–745. https://doi.org/10.1105/tpc.15.00946
15. Lu C, Chen MX, Liu R, Zhang L, Hou X, Liu S, Ding X, Jiang Y, Xu J, Zhang J, Zhao X, Liu YG (2019) Abscisic acid regulates auxin distribution to mediate maize lateral root development under salt stress. Front Plant Sci 10:716. https://doi.org/10.3389/fpls.2019.00716
16. Chen MX, Lu CC, Sun PC, Nie YX, Tian Y, Hu QJ, Das D, Hou XX, Gao B, Chen X, Liu SX, Zheng CC, Zhao XY, Dai L, Zhang J, Liu YG (2021) Comprehensive transcriptome and proteome analyses reveal a novel sodium chloride responsive gene network in maize seed tissues during germination. Plant Cell Environ 44(1): 88–101. https://doi.org/10.1111/pce.13849

Chapter 13

Phosphoproteomic Approaches to Evaluate ABA Signaling

Kota Yamashita and Taishi Umezawa

Abstract

Abscisic acid (ABA) is a major phytohormone that regulates various processes in plants (e.g., seed dormancy/germination, abiotic/biotic stress responses). As protein phosphorylation is involved in the major pathways of ABA signaling, it is necessary to elucidate the phosphosignaling pathway involved in the ABA response. Phosphoproteomics enables determination of the proteins phosphorylated in vivo, and recent studies have applied a comparative phosphoproteomic approach to analyze ABA signaling in plants. For example, ABA-responsive phosphoproteins were identified in barley embryos. Furthermore, a phosphoproteomic approach is useful for screening protein kinase substrates by comparative analysis using kinase knockout mutants. Here, some technical points regarding phosphoproteomic analyses of ABA responses in plants are described.

Key words Abscisic acid, Protein phosphorylation, Signal transduction, Phosphoproteomics, Mass spectrometry, Label-free quantification

1 Introduction

As sessile organisms, plants have developed their own mechanisms to respond to various environmental conditions. Abscisic acid (ABA) is a major phytohormone that has critical roles in abiotic stress responses in plants. In 2009, several studies revealed the major ABA signaling pathway, in which protein phosphorylation and dephosphorylation are employed as key processes. There are three core components: PYRABACTIN RESISTANCE1 (PYR1)/PYR1-LIKE (PYL)/REGULATORY COMPONENTS OF ABA RECEPTOR (RCAR) ABA receptor, protein phosphatase 2C (PP2C), and SNF1-related protein kinase 2 (SnRK2) [1, 2]. In the absence of ABA, PP2Cs interact with and dephosphorylate SnRK2s. Upon binding to ABA, ABA receptors interact with and inhibit PP2Cs, leading to SnRK2 activation [3]. ABA-activated SnRK2s phosphorylate various proteins such as transcription factors, ion channels, and transporters. A number of SnRK2 substrates have been identified by phosphoproteomic approaches [4, 5].

Takuya Yoshida (ed.), *Abscisic Acid: Methods and Protocols*,
Methods in Molecular Biology, vol. 2462, https://doi.org/10.1007/978-1-0716-2156-1_13,

Phosphoproteomics is a technology used to obtain large-scale data of phosphoproteins/peptides [6]. As described above, a phosphoproteomic approach is useful for screening differentially regulated phosphoproteins or protein kinase substrates. For this purpose, the experimental design is critical for obtaining optimal results. First, the selection of the plant samples to evaluate (e.g., wild-type and knockout mutants) is important. It is also important to identify suitable conditions under which protein phosphorylation is significantly changed among samples. For example, the dynamics of kinase activity are critical for identifying protein kinase substrates. Prior to phosphoproteomic analysis, it must be determined when and where protein phosphorylation mediated by the target kinase can be observed.

To date, several methods have been developed for phosphoproteomics [6–8]. An example of one such technique is shot-gun analysis in combination with an efficient method for phosphopeptide enrichment (e.g., metal oxide chromatography, immobilized metal affinity chromatography, and cation exchange chromatography). A single method or a combination of methods can be used. It should be noted that each method is complementary. Shot-gun analysis can be performed using high-performance and high-accuracy liquid chromatography with tandem mass spectrometry (LC-MS/MS). Recently, time-of-flight or Fourier transform MS are widely used in this field.

LC-MS/MS generates complex data, which need to be processed for identification and quantitation of phosphopeptides. Data processing consists of two major steps, that is identification of peptides by a search engine (Subheading 3.3.1) and quantitation by analytical software (Subheadings 3.3.2 and 3.3.3). Several software programs are available for peptide identification, such as Mascot (Matrix Science) [9], SEQUEST (Thermo Fisher Scientific) [10], Andromeda [11], MS-GF+ [12], Comet [13], X!Tandem [14], and MSFragger [15].

There are several methods of MS-based quantification of phosphopeptides. Before starting the experiments, choose a strategy for quantifying MS data. For example, in addition to isotope-labeled or isobaric-tagged approaches, such as stable isotope labeling by amino acids in cell culture, tandem mass tags, and isobaric tags for relative and absolute quantitation [6–8], label-free quantitation is widely used for phosphoproteomics. Although label-free quantification (LFQ) is less accurate, it can be applied without using any specific reagents or sample preparation. In this chapter, we describe an experimental protocol for LFQ.

In LFQ, peptides are quantitated using algorithms based on extracted ion chromatography (XIC) or using spectral counting algorithms based on tandem MS spectra. Some software programs have been developed for XIC-based analysis, including MaxLFQ in MaxQuant [16], Skyline (MacCoss Lab Software) [17], and

FeatureFinderIdentification (*FFId*) in OpenMS [18]. For spectral counting, we can use the normalized spectral index [19], exponentially modified protein abundance index [20], and normalized spectral abundance factor [21]. OpenMS, an open-source and cross platform software, can also be used for XIC analysis. *FFId* is implemented in LFQ methods using OpenMS/KNIME [22].

After LC-MS analysis and data processing, some candidate phosphopeptides will be identified by comparative analysis. The next step is validation of the results. If a responsible protein kinase has already been identified, we will confirm whether the phosphoprotein is an actual substrate of the kinase. Then the following features will be determined: whether the phosphoprotein is involved in signal transduction pathways, whether it is a positive or negative regulator of the biological responses, and how phosphorylation affects protein function.

Such technical development enables the analysis of thousands of phosphoproteins and their in vivo dynamics, thus providing useful information to predict protein phosphorylation networks involved in a variety of signaling pathways [7, 8]. In this chapter, we describe a protocol for phosphoproteomic analysis, including verification by in vitro assay, in our recent study (Fig. 1). In that study, we compared dormant and after-ripened barley seeds and employed LFQ of phosphopeptides enriched by hydroxyl acid-modified metal oxide chromatography (HAMMOC) [23]. The data clearly showed differential regulation of ABA-responsive phosphoproteins between dormant and after-ripened barley seeds.

2 Materials

All solutions are prepared using ultrapure water and reagents of the highest purity or grade.

2.1 Sample Preparation

1. Petri dish.
2. Filter paper.
3. Scalpels or razor blades.
4. Barley grains.
5. 100 mM ABA solution: Dissolve 26.4 mg (±)-ABA in 1 mL of ethanol and store at −20 °C.
6. Tris buffer: 10 mM Tris-HCl, pH 9.0. Store at 4 °C.
7. Extraction buffer (prepare just before use): 8 M urea in Tris buffer, phosphatase inhibitor cocktails (*see* **Note 1**).
8. Liquid nitrogen.
9. Bead mill.
10. Microcentrifuge.

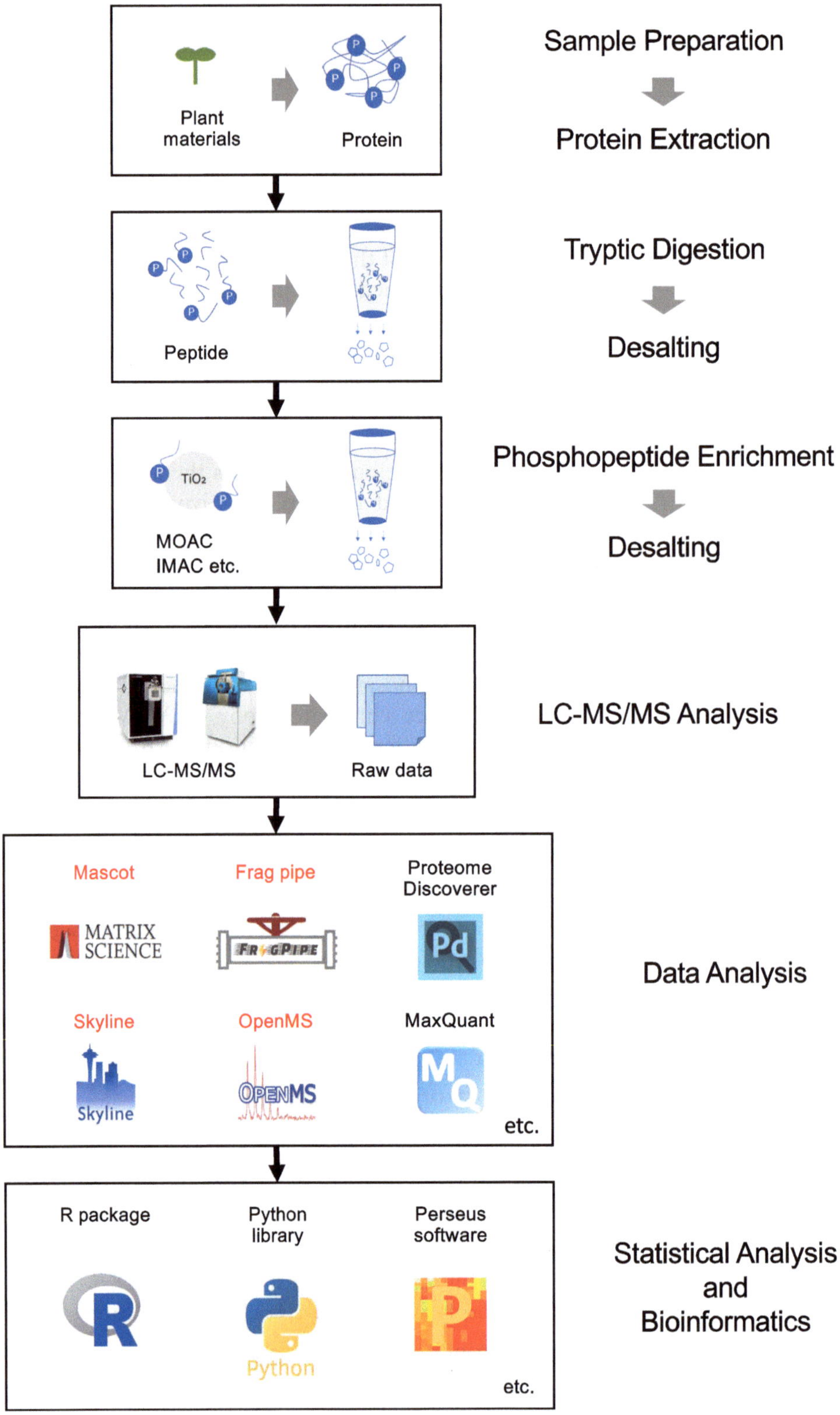

Fig. 1 Overview of phosphoproteomic analysis workflow. Red texts indicate mentioned instrument or software in this chapter. Firstly, extracted protein from any plant materials is digested by trypsin. Phosphopeptides are

11. Microcentrifuge tubes.
12. Protein assay reagent.

2.2 Phosphoproteomic Analysis

1. 50 mM NH_4HCO_3 (prepare just before use): Dissolve 3.95 mg NH_4HCO_3 in 1 mL of water.
2. Reduction buffer (prepare just before use): 10 mM DTT, 50 mM NH_4HCO_3.
3. Alkylation buffer (prepare just before use): 50 mM iodoacetamide, 50 mM NH_4HCO_3.
4. Lysyl endopeptidase (LysC), MS grade.
5. Trypsin, MS grade.
6. 2% (v/v) trifluoroacetic acid (TFA).
7. pH test paper.
8. Solution A: 80% (v/v) acetonitrile (ACN), 0.1% (v/v) TFA.
9. Solution B: 5% (v/v) ACN, 0.1% (v/v) TFA.
10. 15 mL conical tubes.
11. Tweezers or scissors.
12. Empore C18 extraction disk.
13. Empore C8 extraction disk.
14. 16-gauge needle (e.g., Hamilton Kel-F hub (KF) needle) to punch out small disks for preparing 200 μL pipet tip used in Stage-Tip columns.
15. 13-gauge needles (e.g., Hamilton Kel-F hub (KF) needle) to punch out small disks for preparing 1 mL pipet tip used in Stage-Tip columns.
16. Plunger Assemblies (e.g., Hamilton) for the 13- or 16-gauge needles.
17. C18 1 mL Stage-Tip: 1 mL pipette tip packed with four or five C18 disks made with the 13-gauge needle. Can be kept at room temperature.
18. C8 200 μL Stage-Tip: 200 μL pipet tip packed with a single C8 disk made with the 16-gauge needle. Can be kept at room temperature.
19. Titania (TiO_2) beads stored under dry conditions to maintain specificity for phosphopeptides (*see* **Note 2**).

Fig. 1 (continued) enriched by any methods, for example, metal oxide affinity chromatography (MOAC), and Immobilized metal affinity chromatography (IMAC). Enriched phosphopeptides are analyzed by an LC-MS/MS system (AB SCIEX, Thermo, Bruker, etc.). MS raw data is processed by proteomics software, for example, Mascot, Skyline, Frag pipe, OpenMS, and MaxQuant. Finally, normalized peptide or protein abundance are analyzed statistically and bioinformatically by any tools (e.g., R package, Python library, Perseus software)

20. Methanol.
21. Solution C: 300 mg/mL lactic acid in solution A.
22. HAMMOC Stage-Tip: C8 200 μL Stage-Tip with 3 mg TiO_2 beads prepared just before use (*see* **Note 2**).
23. 20% (v/v) Phosphate.
24. 5% (v/v) Ammonia.
25. 10% (v/v) Piperidine.
26. C18 200 μL Stage-Tip: 200 μL pipet tip packed with a single or two C18 disk made with the 16-gauge needle. Can be kept at room temperature.
27. SpeedVac.
28. 0.1% (v/v) Formic acid (FA).
29. Solvent A: 2% (v/v) ACN, 0.1% (v/v) FA.
30. Solvent B: 80% (v/v) ACN, 0.1% (v/v) FA.
31. LC autosampler vial.
32. Capillary HPLC Column (e.g., GL science MonoCap C18 High Resolution 2000 column or equivalent).
33. LC-MS/MS systems consisting of an autosampler (e.g., Eksigent Autosampler-2 1D plus), LC system (e.g., Eksigent NanoLC Ultra), and MS instrument (e.g., AB SCIEX TripleTOF 5600).

2.3 Data Analysis

1. MS data file format converter (e.g., msConvert, AB SCIEX Protein Pilot software).
2. Peptide sequence and protein ID search engine (e.g., Mascot sever, MSFragger).
3. Quantitative proteomics software (e.g., Skyline, Frag pipe, OpenMS).

2.4 Normalization of the Peptide Abundance, Statical Analysis, and Bioinformatics

1. Statical analysis tools (e.g., R package, Python library).
2. Bioinformatics tools (e.g., rmotifx R package, DAVID, https://david.ncifcrf.gov).

2.5 In Vitro Phosphorylation Assay

1. Expression vectors (e.g., pGEX series) in *E. coli*.
2. *E. coli* expression strains (e.g., BL21[DE3]).
3. 2× YT medium: Dissolve 16 g Bacto tryptone, 10 g Bacto yeast extract, and 5 g NaCl in 800 mL of H_2O. Adjust pH to 7.2 with 1N NaOH, and then make up the volume up to 1 L with H_2O. Sterilize by autoclaving. Add antibiotics after cooling.

4. 1 M IPTG: Dissolve 238 mg isopropyl-β-D-thiogalactoside (IPTG) in 10 mL of H_2O. Sterilize by filtration and store at −20 °C.
5. Lysis buffer: 20 mM Tris-HCl (pH 8.0), 1 mM EDTA, 1% TritonX-100, 2.5 mg/mL Lysozyme.
6. Sonicator.
7. 1,4-dithiothreitol (DTT).
8. Rotator.
9. Affinity resins (e.g., Glutathione Sepharose 4B resin).
10. Phosphate buffered saline (PBS): Dissolve 8 g NaCl, 0.2 g KCl, 1.44 g Na_2HPO_4, and 0.24 g KH_2PO_4 in 800 mL of H_2O. Adjust pH to 7.5 with 1N HCl, and then make up the volume to 1 L with H_2O.
11. Elution buffer (e.g., GST-elution buffer: 50 mM Tris-HCl (pH 8.0), 10 mM reduced glutathione).
12. 2× sample loading buffer: For 1 mL, 125 μL of 1 M Tris-HCl (pH 6.8), 40 mg SDS, 100 mg sucrose, 0.04 mg BPB, and 31 mg DTT.
13. SDS-PAGE system.
14. 10× reaction buffer: For 100 μL, mix 50 μL of 1 M Tris-HCl (pH 7.5), 10 μL of 1 M $MgCl_2$, 10 μL of 1 M $MnCl_2$, 0.5 μL of 100 mM ATP, and 29.5 μL of H_2O.
15. [γ-^{32}P]-ATP.
16. Gel dryer.
17. Phosphoimager.

3 Methods

All procedures are performed at room temperature unless otherwise specified.

3.1 Sample Preparation

1. Cut the barley grains in half transversally close to the scutellum.
2. Place the embryo half grains on filter paper supplemented with 50 μM ABA solution (or water included equal volume of ethanol) for ABA treatment. The embryos are separated from the half-cut grains and stored at −80 °C until use.
3. Fifteen embryos are ground in liquid nitrogen using a bead mill.
4. Resuspend the samples in 1 mL of extraction buffer and centrifuged at 17,400 × *g* at 4 °C for 10 min.

5. Transfer the supernatant to a new tube, and measure the protein concentration of the extract. The extract can be stored at −80 °C until use.

3.2 Phosphoproteomic Analysis

3.2.1 Tryptic Digestion

1. Prepare aliquots of 400 μg total protein from the extract (*see* **Note 3**).
2. Add Reduction buffer (8 μL) to the aliquots (1 μL of buffer for every 50 μg protein) and incubate for 30 min at room temperature (25 °C).
3. Add Alkylation buffer (8 μL) to the reduced sample (1 μL of buffer for every 50 μg protein), and incubate in the dark for 20 min at room temperature (25 °C).
4. Add LysC to the alkylated sample (1 μg for every 100 μg protein) and incubate for more than 3 h at room temperature (25 °C).
5. Dilute the LysC digest with 4 volumes of 50 mM NH_4HCO_3.
6. Add Trypsin to the diluted sample (1 μg for every 50 μg protein) and incubate overnight at room temperature (25 °C).

3.2.2 Stage-Tip and Spin Adaptor Preparation

1. Place the Empore C8 or C18 extraction disks on a clean disposable plastic petri dish. The C8 and C18 disks are used for HAMMOC Stage-Tip and desalting Stage-Tip, respectively (*see* **Note 4**).
2. Punch out small disks using needles with proper inner diameters.
3. Push the disk out of the needle and fix it into tapering of a pipet tip using plunger assemblies.
4. Drill a hole in the lids of microcentrifuge tubes or 15 mL conical tubes using a sharp end of a pair of tweezers or scissors. They are used as spin adaptors for the 200 μL or 1 mL Stage-Tips, respectively.

3.2.3 Desalting the Digests Using a 1 mL C18 Stage-Tip (1 mL)

Centrifuge time for each operation is approximate. Continue to spin the C18 1 mL Stage-Tip to force all buffer through.

1. Prepare a C18 1 mL Stage-Tip as describe in Subheading 3.2.2.
2. Place the C18 1 mL Stage-Tip on the spin adaptor, that is, a 15 mL conical tube with a hole in the lid (**step 4**, Subheading 3.2.2).
3. The tryptic digests (Subheading 3.2.1) are acidified with an equal volume of 2% TFA (*see* **Note 5**).
4. Load 1 mL of solution A onto the C18 Stage-Tip for conditioning and centrifuged at 2300 × *g* for 5–10 min.
5. Load 1 mL of solution B onto the C18 Stage-Tip for conditioning and centrifuged at 2300 × *g* for 5–10 min.

6. Load the acidified sample onto the conditioned C18 Stage-Tip and centrifuged at 2300 × *g* for 7–10 min.
7. Load 1 mL of solution B onto the C18 Stage-Tip for washing and centrifuged at 2300 × *g* for 7–10 min.
8. Place the C18 Stage-Tip on a new tube.
9. Load 1 mL of solution A onto the C18 Stage-Tip to elute and centrifuged at 2000 × *g* for 10 min.
10. Dry the eluent in a SpeedVac for 1.5–2.0 h. Dried peptide sample can be stored at −20 °C.

3.2.4 HAMMOC Stage-Tip Column Preparation

1. Prepare a C8 200 μL Stage-Tip as describe in Subheading 3.2.2. The membrane disk is only for keeping TiO_2 beads.
2. Weigh TiO_2 beads and add to the disk in the C8 200 μL Stage-Tip (*see* **Note 6**). To prepare many columns, TiO_2 beads can be suspended in methanol to load on C8 200 μL Stage-Tips. Most importantly, TiO_2 beads have to be kept under dry conditions to keep specificity against phosphopeptides (*see* **Note 2**).

3.2.5 Phosphopeptide Enrichment by HAMMOC

Centrifuge time for each operation is approximate. Continue to spin the C8 200 μL Stage-Tip to force all buffer through.

1. Resuspend the desalted peptides (Subheading 3.2.3) in 200 μL of solution C.
2. Prepare a HAMMOC Stage-Tip as describe in Subheading 3.2.4.
3. Load 80 μL of Solution A onto a HAMMOC Stage-Tip for conditioning, and spin down at 1000 × *g* for 3 min.
4. Load 80 μL of Solution C onto the HAMMOC Stage-Tip for conditioning, and spin down at 1000 × *g* for 3 min.
5. Repeat the **step 4**.
6. Place the HAMMOC Stage-Tip on a new tube.
7. Load the sample onto the HAMMOC Stage-Tip, and spin down at 1000 × *g* for 5 min.
8. Collect the flow-through and load onto the same HAMMOC Stage-Tip again, and spin down at 1000 × *g* for 5 min.
9. Load 80 μL of Solution C onto the HAMMOC Stage-Tip for washing, and spin down at 1000 × *g* for 3 min.
10. Load 80 μL of Solution A onto the HAMMOC Stage-Tip for washing, and spin down at 1000 × *g* for 3 min.
11. Place the HAMMOC Stage-Tip on a tube containing 100 μL of 20% phosphate.
12. Load 50 μL of 5% Ammonia to elute, and spin down at 900 × *g* for 3 min.

13. Load 50 μL of 10% Piperidine to elute, and spin down at 900 × *g* for 3 min.

3.2.6 Desalting Phosphopeptides Using a 200 L C18 Stage-Tip

Centrifuge time for each operation is approximate. Continue to spin the C18 200 μL Stage-Tip to force all buffer through.

1. Prepare a C18 200 μL Stage-Tip as describe in Subheading 3.2.2.
2. Load 40 μL of Solution A onto the C18 Stage-Tip for conditioning, and spin down at 1000 × *g* for 2 min.
3. Load 40 μL of Solution B onto the C18 Stage-Tip for conditioning, and spin down at 1000 × *g* for 2 min.
4. Load the sample (Subheading 3.2.5) onto the C18 Stage-Tip, and spin down at 1000 × *g* for 5 min.
5. Load 40 μL of Solution B onto the C18 Stage-Tip for washing, and spin down at 1000 × *g* for 2 min.
6. Place the C18 Stage-Tip on a new tube.
7. Load 50 μL of Solution A onto the C18 Stage-Tip to elute, and spin down at 900 × *g* for 3 min.
8. Dry the eluent in a SpeedVac for 15–20 min. Dried peptide sample can be stored at −20 °C.

3.2.7 LC-MS/MS Analysis

The LC-MS/MS instrument is prepared. Our method used to prepare the TripleTOF 5600 system is shown below as an example.

1. Create a method for the LC-MS/MS instrument (*see* **Note** 7).
2. Resuspend the desalted peptides in 20 μL of 0.1% (v/v) FA and transfer a vial for LC autosampler.
3. Place the samples in an LC autosampler cooled to 4–8 °C.
4. Set up the nano-spray column with a column heater set to 50 °C.
5. Analyze each phosphopeptide sample using the HPLC and MS parameters set in **step 1**. For each analysis, we typically load 6–10 μL of sample at a constant pressure of 500 nL/min.

3.3 Data Analysis: Peptide Identification and Quantitative Analysis

Data analysis consists of two major steps: peptide identification and quantitative analysis. In Subheading 3.3.1, we describe peptide identification using the Mascot sever (version 2.4.1) or MSFragger (version 3.1.1) with Frag Pipe (version 14.0). For the quantitative analysis, we describe the XIC-based LFQ method using the Skyline (version 20.2) (Subheading 3.3.2) or *FeatureFinderIdentification* (*FFId*) in OpenMS (version 2.60) (Subheading 3.3.3).

3.3.1 Database Search to Identify Peptides and Phosphorylation Sites

1. For the Mascot search, MS raw data will be converted, from a wiff/wiff.scan file derived from the TripleTOF5600 system to Mascot generic format (MGF) file by the vendor software (e.g.,

AB SCIEX Protein Pilot). For the MSFragger search, MS raw data is converted to mzML file as open format by msConvert (*see* **Note 8**).

2. Set the search parameters for Mascot or MSFragger (*see* **Note 9**).
3. Load the MGF file into Mascot server with Mascot Daemon that automates the submission of data file to Mascot server (*see* **Note 10**). Load mzML file into MSFragger with Frag pipe, name the experiments, and set the replicates (*see* **Note 11**).
4. Search peptide sequence and protein ID with parameter settings in **step 3**.

3.3.2 XIC-Based Label-Free Quantification (LFQ) of Phosphopeptides Using the Skyline Software

1. Load result files of peptide identification, for example, Mascot DAT file, to Skyline software and generate a spectral library (BLIB file) (*see* **Note 12**).
2. Load MS raw data, and set the MS1 full-scan parameter which depends on MS instruments (*see* **Note 13**).
3. Extract precursor ion chromatogram from MS raw data based on the peptide identifications.
4. To improve quantitative accuracy, reintegrate the XIC using mProphet algorithm.

3.3.3 XIC-Based LFQ Method Using the FFId in OpenMS Software

1. Convert the result files of peptide identification, for example, PEPXML file from MSFragger, to idXML file using *IDFileConverter* (*see* **Notes 14** and **15**).
2. Annotate target or decoy discrimination to ID files using *PeptideIndexer.*
3. Select peptide IDs with PSM level FDR < 1% using *FalseDiscoveryRate.*
4. To minimize the retention time (RT) difference derived among multiple LC-MSMS runs, align the RT of ID files using *MapAlignnerIdentification* with the "linear" smoothing option.
5. Scale the original RT of mzML files to the aligned RT using *MapRTtoransformer* with the "linear" smoothing option.
6. Detect a peptide-specific signal patterns, called features, in the scaled mzML files using *FFId* (*see* **Note 16**). For support vector machine classification implemented in *FFId*, the ID file derived from same mzML file is set to the "internal IDs" input. Subsequently, all ID files are merged using *IDMerger* and set to "external IDs" input, respectively.

1. Normalize quantitative data from the XIC peak areas (*see* **Note 17**).

3.4 Normalization of the Peptide Abundance, Statical Analysis, and Bioinformatics

2. Perform statistical analyses using R package, Python library, and other tools.
3. Filter out the unlocalized phosphorylation sites using a localization probability of 75% as the cut-off.
4. For motif analysis, 13 amino acids around phosphorylated serine/threonine/tyrosine residues are extracted from identified phosphopeptide sequences and analyzed using the Motif-X algorithm in the rmotifx R package.
5. For gene ontology (GO) analysis, load the annotated data to DAVID. The result can be visualized using REVIGO.

3.5 In Vitro Phosphorylation Assay

1. Prepare plasmid constructs in which a cDNA of the protein kinase gene is cloned into an appropriate expression vector.
2. Prepare plasmid constructs in which the cDNA of a substrate gene fragment is cloned into an expression vector. A short fragment or synthetic peptide can be used containing phosphorylation site(s) when such information is known.
3. Introduce a mutation to the phosphorylation site in the substrate gene by site-directed mutagenesis; for example, Ser/Thr should be changed to Ala. That mutation enables confirmation of the phosphorylation site.
4. After transformation of *E. coli*, the *E. coli* cells carrying expression plasmids are cultured at 37 °C, and protein expression is induced by the addition of IPTG to 0.1–1 mM at OD_{600} 0.5–0.6 in 2 × YT medium.
5. After about 10 h incubation at 15 °C, harvested the cells by 4000 × *g* centrifugation for 6 min at 4 °C and the pellet is resuspended in the ice-cold lysis buffer.
6. Sonicate the cells on ice four times for 20 s with regular resting intervals.
7. After sonication, add DTT to 18 mM and the cells are subsequently rotated for 30 min at 4 °C.
8. Obtain a supernatant by centrifugation at 17,400 × *g* for 15 min at 4 °C.
9. For the affinity purification of any tagged recombinant proteins (e.g., GST-tagged), coincubate the supernatant with an appropriate affinity resin (e.g., Glutathione Sepharose 4B resin) for 30 min at 4 °C.
10. Wash the resin four times in the PBS.
11. Resuspend the washed resin with any an ice-cold elution buffer (e.g., GST-elution buffer) for a few minutes.
12. Measure the protein concentration.

13. Prepare the reaction mix on ice by mixing 0.1–1 μg protein kinase, 1 μg substrate proteins, 1 μL of 10× reaction buffer, and 0.5 μL of [γ-^{32}P]-ATP. Then make up the volume to 10 μL with water. For multiple reactions, a premix solution without proteins can be made.
14. Incubate the reaction mix at 30 °C for 30 min (*see* **Note 18**).
15. Mix 10 μL of aliquots with 10 μL of 2× sample loading buffer and resolved by SDS-PAGE.
16. Wash the gel twice in 100 mL of water at room temperature for 10 min.
17. Drying the gel by gel dryer.
18. Detect radioisotope activity by autoradiography.

4 Notes

1. It is necessary to optimize the buffer composition depending on the sample type. Factors such as tissue type(s), cellular localization, and plant species should be taken into consideration. Phosphatase inhibitors may not be required in the denaturing extraction buffer. If alternative extraction buffers without denaturants are used, phosphatase inhibitors are needed.
2. The specificity of TiO_2 beads for phosphopeptides is reduced by water absorption when stored without desiccation. Specificity can be recovered by heating the beads in a drying oven at 130 °C for 30 min.
3. The amount of protein should be optimized in advance. The sample should not be heated above 37 °C to avoid artificial protein modification by a urea derivative.
4. Commercially available tips can be used.
5. Acidification is required for peptide binding to the C18 membrane disks. Check the pH of the sample with a pH test paper and add TFA until it reaches pH 2–3.
6. We usually use 3 mg TiO_2 beads for a single column and apply 400 μg of protein digests. Specificity and capacity of TiO_2 beads depend on samples.
7. The parameter setting for HPLC is shown in Table 1. To identify more peptides by MS, we use a gradient program in Table 2. The parameter setting for MS is shown in Table 3.
8. The parameter settings for Mascot or MSFragger in our phosphoproteomic analysis are shown in Table 4.
9. The tutorial of Mascot server and Mascot daemon can be found at http://www.matrixscience.com/mascot_support.html.

Table 1
Parameter setting of HPLC

Autosampler-2 1D plus (Eksigent), NanoLC Ultra (Eksigent)	
Column	MonoCap C18 High Resolution 2000 column (GL science)
Solvent A	2% (v/v) ACN, 0.1% (v/v) FA
Solvent B	80% (v/v) ACN, 0.1% (v/v) FA
Flow rate	500 nL/min on a 360 min gradient
Oven temp	50 °C

Table 2
Gradient program

Step	Solvent A (%)	Solvent B (%)	Time (min)
First	98	2	0
Second	60	40	300
Third	10	90	20
Fourth	2	98	40
Total			360

Table 3
The setting of TripleTOF 5600 system

Parameter	Setting
Ion spray voltage	2300 V
Scan mode	Positive
Full MS scan (TOF-MS)	
Scan range	*m/z* 400–1250
Accumulation time	250 ms
DDA-MS/MS scan (TOF-MS)	
Optimization	High sensitivity mode
Collision energy	37.0 V
Scan range	*m/z* 100–1600
Accumulation time	50 ms
Total cycle time	1.3 s
Scan spectral rate	20 Hz
Dynamic exclusion	10 s

Table 4
Peptide sequence and protein ID search parameter

Parameter	Settings
Precursor mass tolerance	±50 ppm
Fragment ion mass tolerance	±0.05 Da
Precursor ion charge	+2, +3, +4
Digestion enzyme	Trypsin
Allowed missed cleavages	2
Fixed modification	Carbamidomethylation (C)
Variable modification	Oxidation (M), Acetylation (N-term), Phosphorylation (STY)
Allowed false discovery rate	<1%
Peptide length	7–50
Peptide mass range	400–5000
Max variable mods on a peptide	3

10. The tutorial of msConvert for MSFragger can be found at http://msfragger.nesvilab.org/tutorial_convert.html.
11. The tutorial of MSFragger and Frag pipe can be found at https://msfragger.nesvilab.org/tutorial_fragpipe.html.
12. The tutorial of Skyline called "MS1 Full-Scan Filtering" can be found at https://skyline.ms/wiki/home/software/Skyline/page.view?name=tutorials.
13. For MS1 filtering, we set the precursor mass analyzer to "TOF," the resolving power to 20,000, and the retention time filtering to "include all matching scans."
14. The tutorial of OpenMS can be found at https://www.openms.de/tutorials/.
15. If you enable PeptideProphet [24] or PTMProphet [25] parameter in Frag pipe, PEPXML file is prefixed "intract-" or "intract-mod". These files can also be used as peptide identification files, respectively.
16. We set the *m/z* window size and the retention time window size for chromatogram extraction of *FFId* to 50 ppm and 120 s, respectively.
17. In our methods, the XIC peak areas will be normalized using the Levenberg–Marquardt optimization [16]. The R script is available from GitHub (https://github.com/Ume-lab/raf22 raf36_phosphoproteome).
18. Incubation should not be longer than 30 min, as this can decrease phosphorylation.

Acknowledgments

This work was supported by the Japan Science and Technology Agency program PRESTO and Grants-in-Aid for Scientific Research from the Ministry of Education, Culture, Sports, Science and Technology, Japan.

References

1. Cutler SR, Rodriguez PL, Finkelstein RR, Abrams SR (2010) Abscisic acid: emergence of a core signaling network. Annu Rev Plant Biol 61:651–679. https://doi.org/10.1146/annurev-arplant-042809-112122
2. Umezawa T, Nakashima K, Miyakawa T et al (2010) Molecular basis of the core regulatory network in abscisic acid responses: sensing, signaling, and transport. Plant Cell Physiol 51: 1821–1839. https://doi.org/10.1093/pcp/pcq156
3. Umezawa T, Sugiyama N, Mizoguchi M et al (2009) Type 2C protein phosphatases directly regulate abscisic acid-activated protein kinases in Arabidopsis. Proc Natl Acad Sci U S A 106: 17588–17593. https://doi.org/10.1073/pnas.0907095106
4. Umezawa T, Sugiyama N, Takahashi F et al (2013) Genetics and phosphoproteomics reveal a protein phosphorylation network in the abscisic acid signaling pathway in Arabidopsis thaliana. Sci Signal 6:rs8. https://doi.org/10.1126/scisignal.2003509
5. Wang P, Xue L, Batelli G et al (2013) Quantitative phosphoproteomics identifies SnRK2 protein kinase substrates and reveals the effectors of abscisic acid action. Proc Natl Acad Sci 110:11205–11210. https://doi.org/10.1073/pnas.1308974110
6. Macek B, Mann M, Olsen JV (2009) Global and site-specific quantitative phosphoproteomics: principles and applications. Annu Rev Pharmacol Toxicol 49:199–221. https://doi.org/10.1146/annurev.pharmtox.011008.145606
7. de la Fuente van Bentem S, Hirt H (2007) Using phosphoproteomics to reveal signalling dynamics in plants. Trends Plant Sci 12:404–411. https://doi.org/10.1016/j.tplants.2007.08.007
8. Peck SC (2006) Phosphoproteomics in Arabidopsis: moving from empirical to predictive science. J Exp Bot 57:1523–1527. https://doi.org/10.1093/jxb/erj126
9. Koenig T, Menze BH, Kirchner M et al (2008) Robust prediction of the MASCOT score for an improved quality assessment in mass spectrometric proteomics. J Proteome Res 7:3708–3717. https://doi.org/10.1021/pr700859x
10. Eng JK, McCormack AL, Yates JR (1994) An approach to correlate tandem mass spectral data of peptides with amino acid sequences in a protein database. J Am Soc Mass Spectrom 5: 976–989. https://doi.org/10.1016/1044-0305(94)80016-2
11. Cox J, Neuhauser N, Michalski A et al (2011) Andromeda: a peptide search engine integrated into the MaxQuant environment. J Proteome Res 10:1794–1805. https://doi.org/10.1021/pr101065j
12. Kim S, Pevzner PA (2014) MS-GF+ makes progress towards a universal database search tool for proteomics. Nat Commun 5:5277. https://doi.org/10.1038/ncomms6277
13. Eng JK, Jahan TA, Hoopmann MR (2013) Comet: an open-source MS/MS sequence database search tool. Proteomics 13:22–24. https://doi.org/10.1002/pmic.201200439
14. Craig R, Beavis RC (2004) TANDEM: matching proteins with tandem mass spectra. Bioinformatics 20:1466–1467. https://doi.org/10.1093/bioinformatics/bth092
15. Kong AT, Leprevost FV, Avtonomov DM et al (2017) MSFragger: ultrafast and comprehensive peptide identification in mass spectrometry–based proteomics. Nat Methods 14:513–520. https://doi.org/10.1038/nmeth.4256
16. Cox J, Hein MY, Luber CA et al (2014) Accurate proteome-wide label-free quantification by delayed normalization and maximal peptide ratio extraction, termed MaxLFQ. Mol Cell Proteom 13:2513–2526. https://doi.org/10.1074/mcp.M113.031591
17. Schilling B, Rardin MJ, MacLean BX et al (2012) Platform-independent and label-free quantitation of proteomic data using MS1 extracted ion chromatograms in skyline application to protein acetylation and phosphorylation. Mol Cell Proteomics 11:202–214. https://doi.org/10.1074/mcp.M112.017707

18. Weisser H, Choudhary JS (2017) Targeted feature detection for data-dependent shotgun proteomics. J Proteome Res 16:2964–2974. https://doi.org/10.1021/acs.jproteome.7b00248
19. Griffin NM, Yu J, Long F et al (2010) Label-free, normalized quantification of complex mass spectrometry data for proteomic analysis. Nat Biotechnol 28:83–89. https://doi.org/10.1038/nbt.1592
20. Ishihama Y, Oda Y, Tabata T et al (2005) Exponentially modified protein abundance index (emPAI) for estimation of absolute protein amount in proteomics by the number of sequenced peptides per protein*S. Mol Cell Proteomics 4:1265–1272. https://doi.org/10.1074/mcp.M500061-MCP200
21. Zybailov B, Mosley AL, Sardiu ME et al (2006) Statistical analysis of membrane proteome expression changes in Saccharomyces cerevisiae. J Proteome Res 5:2339–2347. https://doi.org/10.1021/pr060161n
22. Röst HL, Sachsenberg T, Aiche S et al (2016) OpenMS: a flexible open-source software platform for mass spectrometry data analysis. Nat Methods 13:741–748. https://doi.org/10.1038/nmeth.3959
23. Ishikawa S, Barrero JM, Takahashi F et al (2019) Comparative phosphoproteomic analysis reveals a decay of ABA signaling in barley embryos during after-ripening. Plant Cell Physiol 60:2758–2768. https://doi.org/10.1093/pcp/pcz163
24. Ma K, Vitek O, Nesvizhskii AI (2012) A statistical model-building perspective to identification of MS/MS spectra with PeptideProphet. BMC Bioinformatics 13:S1. https://doi.org/10.1186/1471-2105-13-S16-S1
25. Shteynberg DD, Deutsch EW, Campbell DS et al (2019) PTMProphet: fast and accurate mass modification localization for the trans-proteomic pipeline. J Proteome Res 18:4262–4272. https://doi.org/10.1021/acs.jproteome.9b00205

Chapter 14

Affinity Purification Followed by Liquid Chromatography–Tandem Mass Spectrometry to Identify Proteins Interacting with ABA Signaling Components

Fumiyuki Soma, Fuminori Takahashi, Kazuo Shinozaki, and Kazuko Yamaguchi-Shinozaki

Abstract

Abscisic acid (ABA) is a key phytohormone involved in plant development, seed germination and responses to osmotic stresses, such as drought and high salinity. SNF1-related protein kinases (SnRK2s) play important roles in ABA-dependent and ABA-independent osmotic stress signaling. SnRK2s phosphorylate transcription factors and ion channels in response to ABA or osmotic stress to induce the expression of stress-responsive genes and stomatal closure, respectively, to confer osmotic stress tolerance. The activity of SnRK2s is directly or indirectly regulated by several protein factors. Identification of downstream substrates or upstream regulators of SnRK2s is very useful for elucidating protein components that regulate ABA and osmotic stress signaling. Here, we describe the use of affinity purification by coimmunoprecipitation and liquid chromatography–tandem mass spectrometry to identify protein complexes involved in ABA and osmotic stress signaling in plants. We previously identified several protein factors that regulate ABA and osmotic stress signaling by using this method.

Key words GFP-tagged protein, Coimmunoprecipitation, In-gel digestion, Peptide purification, LC-MS/MS analysis

1 Introduction

Osmotic stresses, such as drought stress, are adverse environmental factors for plant growth. Abscisic acid (ABA) is a key hormone for plant growth, development and osmotic stress responses [1]. Many protein components have been reported to participate in these processes [2]. Among them, subclass III SNF1-related protein kinases (SnRK2s), protein phosphatase 2Cs (PP2Cs) and ABA receptors PYRABACTIN RESISTANCE1/PYR1-LIKE/REGULATORY COMPONENT OF ABA RECEPTORS (PYR/PYL/RCAR) are known as core ABA signaling components. Under nonstress conditions, PP2Cs dephosphorylate and repress

Takuya Yoshida (ed.), *Abscisic Acid: Methods and Protocols*,
Methods in Molecular Biology, vol. 2462, https://doi.org/10.1007/978-1-0716-2156-1_14,

SnRK2s. The ABA-binding proteins PYR/PYL/RCAR inhibit PP2Cs, resulting in activation of SnRK2s under osmotic stress conditions [3]. Activated subclass III SnRK2s interact with and phosphorylate the bZIP transcription factors ABA responsive element (ABRE)-binding proteins/ABRE binding factors (AREB/ABFs) to induce the expression of stress-responsive genes.

As described above, ABA and osmotic stress signaling is regulated by protein-protein interactions. To understand ABA and osmotic stress signaling, it is important to identify protein factors that interact with and regulate core ABA components. Previously, PYR/PYL/RCAR family proteins were identified as major in vivo interaction partners of ABA INSENSITIVE 1 (ABI1), a PP2C, by using combined coimmunoprecipitation (Co-IP) and liquid chromatography–mass spectrometry (LC-MS/MS) analysis [4]. This method has the advantage of identifying real interacting proteins in plant cells.

Subclass I SnRK2s are activated by osmotic stress in an ABA-independent manner, but the phosphorylation substrates and upstream regulators of subclass I SnRK2s have not been identified. We previously attempted Co-IP and LC-MS/MS analysis by using transgenic Arabidopsis expressing a GFP-tagged subclass I SnRK2 to elucidate osmotic stress signaling mediated by subclass I SnRK2s, and we identified VARICOSE (VCS), an mRNA decapping activator, as a main phosphorylation target of the kinases [5]. In addition, we also identified the kinases RAF18, RAF20 and RAF24 as upstream activators of subclass I SnRK2s by Co-IP and LC-MS/MS analysis [6]. With this interactome analysis as a starting point, we revealed that the RAF18/20/24-subclass I SnRK2-VCS signaling module regulates the mRNA population to maintain plant growth under osmotic stress conditions. By revealing the protein-protein interaction network by Co-IP and LC-MS/MS analysis, we elucidated a novel osmotic stress signaling pathway.

In this chapter, we describe the approach of affinity purification by Co-IP and LC-MS/MS to identify protein complexes involved in ABA and osmotic stress signaling in plants. In this method, we used an in-gel digestion protocol [7] to digest proteins for LC-MS/MS analysis. As interacting proteins identified by using the Co-IP and LC-MS/MS approach may include many indirect targets or false positives, we need to check the interactions by other methods, such as a yeast two-hybrid assay or a bimolecular fluorescence complementation (BiFC) assay.

2 Materials

2.1 Plant Growth and Protein Extraction

1. Transgenic Arabidopsis plants expressing GFP-tagged protein (*see* **Note 1**).

2. Germination medium (GM): 4.6 g/L Murashige and Skoog plant salt mixture, 1 mL/L Gamborg's vitamin solution, 30 g/L sucrose, 8.3 g/L agar (e.g., BD Biosciences Bacto Agar) (*see* **Note 2**), and 0.5 g/L 2-morpholinoethanesulfonic acid monohydrate (MES), adjusted to pH 5.7 with 1 M KOH.
3. Protein extraction buffer: 20 mM HEPES–KOH (pH 7.6), 100 mM NaCl, 0.1 mM EDTA, 5 mM $MgCl_2$, 20% (v/v) glycerol, 0.5% (v/v) Triton X-100.
4. 50 mL tube.
5. Mortar and pestle.
6. Tube rotator.
7. Centrifuge.
8. Elution buffer: 50 mM Tris-HCl (pH 6.8), 50 mM dithiothreitol (DTT), 1% (w/v) SDS, 1 mM EDTA, 0.005% (w/v) bromophenol blue, 10% (v/v) glycerol.
9. Filtration material (e.g., Miracloth or equivalent).

2.2 Purification of GFP-Tagged Protein

1. Affinity purification kit for GFP-tagged protein (e.g., Miltenyi Biotec) (*see* **Note 3**).
2. Wash buffer: 20 mM Tris-HCl (pH 7.5).

2.3 SDS-PAGE and In-Gel Digestion

1. Precast gel (e.g., DRC Perfect NT Gel M).
2. Running buffer: 250 mM Tris, 1.92 M glycine, 1% (w/v) SDS.
3. Staining solution: 50% (v/v) methanol, 5% (v/v) acetic acid, 0.25% (w/v) Coomassie Brilliant Blue R-250 (CBB).
4. Shaker and container.
5. Destaining solution: 25% (v/v) methanol, 7.5% (v/v) acetic acid.
6. Ultrapure water (e.g., MilliQ water).
7. Scalpel.
8. 1.5 mL tube.
9. Vacuum centrifuge concentrator.
10. Vortex incubator.
11. Heating block.

2.4 Peptide Purification

1. Acetonitrile (HPLC grade).
2. 25 mM ammonium bicarbonate.
3. DTT solution: 10 mM DTT, 25 mM ammonium bicarbonate.
4. IAA solution: 55 mM iodoacetamide (IAA), 25 mM ammonium bicarbonate.

5. Acetonitrile solution: 50% (v/v) acetonitrile, 12.5 mM ammonium bicarbonate.
6. 1 μg/μL Trypsin/Lys-C protease mix, Mass Spec Grade.
7. Formic acid solution: 50% (v/v) acetonitrile, 5% (v/v) formic acid.
8. Peptide desalting and enrichment chips (e.g., GL Sciences GL-Tip GC and GL-Tip SDB or equivalent) (*see* **Note 4**).
9. Solution A: 5% (v/v) acetonitrile, 0.1% (v/v) TFA.
10. Solution B: 80% (v/v) acetonitrile, 0.1% (v/v) TFA.
11. Centrifuge adaptors.

2.5 LC-MS/MS Analysis

1. LC system (e.g., Eksigent NanoLC Ultra or equivalent) with an autosampler (e.g., Eksigent).
2. Concentration column (column length 5 mm × inner diameter 0.3 mm, particle diameter 5 μm) (e.g., Chemical Evaluation and Research Organization).
3. C18 analytical column (column length 2000 mm × inner diameter 0.1 mm, particle diameter 2 μm) (e.g., GL Sciences MonoCap C18 High Resolution 2000 or equivalent).
4. Spray tip (inner diameter 20 μm, tip inner diameter 10 μm) (e.g., New Objective Inc. PicoTip Emitter SilicaTip or equivalent).
5. 0.1% (v/v) formic acid solution.
6. Solvent A: 2% (v/v) acetonitrile, 0.1% (v/v) formic acid.
7. Solvent B: 80% (v/v) acetonitrile, 0.1% (v/v) formic acid.
8. MS instrument (e.g., SCIEX Triple TOF 5600 system or equivalent).
9. Software (e.g., SCIEX ProteinPilot or equivalent).
10. Protein sequence data by The Arabidopsis Information Resource (TAIR).

3 Methods

3.1 Preparation of Samples

1. Transgenic Arabidopsis plants expressing GFP-tagged protein are used (*see* **Note 1**). Transgenic Arabidopsis plants expressing GFP alone are used as a negative control.
2. Surface-sterilized seeds are sown on GM agar plates. The plates are placed at 4 °C under dark conditions for 3 days and then grown at 22 °C under 16–8-h light-dark cycles for 3 weeks (*see* **Note 5**).

3.2 Protein Extraction

1. A 3.0 g plant sample (fresh weight) is used. Add 9 mL of protein extraction buffer to a 50 mL tube. Plant tissue is ground well in liquid nitrogen, and the ground samples are transferred to the 50 mL tube.
2. The sample is rotated in a 50 mL tube at 4 °C for 30 min (crude protein sample).
3. The crude protein sample is centrifuged at 15,000 × *g* for 10 min at 4 °C, and the supernatant is collected. The supernatant is passed through filtration material (i.e., one layer of Miracloth) and centrifuged at 15,000 × *g* for 20 min at 4 °C, and the supernatant obtained is collected.

3.3 Purification of GFP-Tagged Protein

Here we describe a typical procedure using a commercially available kit (*see* **Note 3**). Please modify and adjust the following steps according to the manufacturer's instructions.

1. Add 80 μL of anti-GFP antibody beads to the crude protein sample, and the sample is incubated for 30 min on ice.
2. M Column is placed in the magnetic field of a μMACS separator at 4 °C, and 200 μL of protein extraction buffer is applied to the column.
3. Elution buffer is prepared in a fresh tube and placed in a preheated 95 °C block.
4. The crude protein sample (**step 1**) is applied to the column.
5. The column is rinsed with 3 × 3 mL of protein extraction buffer.
6. The column is rinsed with 200 μL of wash buffer.
7. Apply 20 μL of preheated 95 °C elution buffer to the column, and the column is incubated for 5 min at room temperature.
8. Apply 50 μL of preheated 95 °C elution buffer to the column, and the eluate (immunoprecipitate) is collected (*see* **Note 6**).

3.4 SDS-PAGE

1. The immunoprecipitated is loaded onto a precast gel, which is run according to the following conditions.
   Run conditions: 20 mA, 60–90 min.
2. After electrophoresis is complete, the gel is stained in CBB staining solution for 1 h at room temperature with gentle shaking.
3. The gel is destained with destaining solution for 6 × 10 min with gentle shaking.
4. The gel is washed with ultrapure water for 3 × 10 min with gentle shaking.
5. The gel is incubated in ultrapure water with gentle shaking overnight.

3.5 In-Gel Digestion

1. Bands of interest or the gel is cut to obtain three parts per lane with a clean scalpel.
2. The excised bands are cut, or a notch is made in the bands (1 mm × 1 mm).
3. Each gel piece is transferred into a fresh 1.5 mL tube.
4. Add 300 μL of acetonitrile per tube, and the tubes are vortexed for 5 min at room temperature. The supernatant is discarded. This process is repeated twice.
5. Evaporation by vacuum centrifugation is performed for 15 min at room temperature.
6. Add 300 μL of DTT solution, and the tube is vortexed at 56 °C for 1 h. The supernatant is discarded.
7. Add 300 μL of 25 mM ammonium bicarbonate, and the tube is vortexed at room temperature for 15 min. The supernatant is discarded.
8. Add 300 μL of IAA solution, and the tube is vortexed at room temperature for 45 min in the dark. The supernatant is discarded.
9. Add 300 μL of 25 mM ammonium bicarbonate, and the tube is vortexed at room temperature for 15 min. The supernatant is discarded.
10. Add 600 μL of acetonitrile solution, and the tube is vortexed at room temperature for 15 min. The supernatant is discarded. This process is reported twice.
11. Evaporation by vacuum centrifugation is performed for 15 min at room temperature.
12. An aliquot of the 1 μg/μL Trypsin/Lys-C Mix is diluted in 25 mM ammonium bicarbonate to obtain a final concentration of 20 ng/μL.
13. Add 150 μL of the Trypsin/Lys-C Mix (20 ng/μL), and the solution is incubated at 37 °C overnight (12–16 h).
14. Add 150 μL of formic acid solution, and the tube is vortexed at room temperature for 30 min. The supernatant is collected. This process is repeated twice, and the supernatants are pooled.
15. Evaporation is performed by vacuum centrifugation until approximately 10 μL remained (crude peptide solution).

3.6 Peptide Purification

Here we describe a typical procedure using commercially available peptide desalting and enrichment chips (*see* **Note 4**). Please modify and adjust the following steps according to the manufacturer's instructions.

1. The crude peptide solution is diluted by adding 40 μL of Solution A.

2. A centrifuge adaptor is connected to a 1.5 mL empty tube, and a GL-Tip SDB column is inserted.
3. Add 20 μL of Solution B, and the column is centrifuged at 3000 × *g* for 2 min at room temperature.
4. Add 20 μL of Solution A, and the column is centrifuged at 3000 × *g* for 2 min at room temperature.
5. The GL-Tip SDB column is placed into a fresh 1.5 mL tube.
6. The crude peptide solution (**step 1**) is added to the GL-Tip SDB column, and the column is centrifuged at 3000 × *g* for 5 min at room temperature. The flow through solution is retained.
7. The GL-Tip SDB column is placed into a 1.5 mL tube.
8. Add 20 μL of Solution A, and the column is centrifuged at 3000 × *g* for 2 min at room temperature.
9. The GL-Tip SDB column is placed into a fresh 1.5 mL tube.
10. Add 50 μL of Solution B, and the column is centrifuged at 3000 × *g* for 3 min at room temperature. The eluate is collected as purified peptide solution A.
11. A centrifuge adaptor is connected to a 1.5 mL empty tube, and a GL-Tip GC column is inserted.
12. Add 20 μL of Solution B, and the column is centrifuged 3000 × *g* for 2 min at room temperature.
13. Add 20 μL of Solution A, and the column is centrifuged at 3000 × *g* for 2 min at room temperature.
14. The GL-Tip GC column is placed into a fresh 1.5 mL tube.
15. The flow through solution (**step 6**) is added to the GL-Tip GC column, and the column is centrifuged at 3000 × *g* for 5 min at room temperature.
16. The GL-Tip GC column is placed into a 1.5 mL tube.
17. Add 20 μL of Solution A, and the column is centrifuged at 3000 × *g* for 2 min at room temperature.
18. The GL-Tip GC column is placed into a fresh 1.5 mL tube.
19. Add 50 μL of Solution B, and the column is centrifuged at 3000 × *g* for 3 min at room temperature. The eluate is collected as purified peptide solution B.
20. Purified peptide solution B (**step 19**) is added to the tube containing peptide solution A (**step 10**). Evaporation is performed by vacuum centrifugation until approximately 10 μL remained (purified peptide solution).

3.7 Identification and Quantification of Proteins by a Nano LC-MS/MS Analyser

1. Samples are concentrated and separated through a concentration column and a separation column by a nano LC system with an autosampler (*see* **Notes 7** and **8**).
2. Ionized samples are measured with a high resolution mass spectrometer (*see* **Note 9**).
3. The MS/MS data are analysed using appropriate software, such as the Paragon algorithm software, and subsequently annotated using the protein sequence data provided by TAIR for peptide identification (*see* **Note 10**).

4 Notes

1. 35Spro:SnRK2-sGFP-expressing Arabidopsis plants etc. can be used [5, 8].
2. We usually use Bacto Agar because plants grow well using this agar. Any other agars can be used.
3. We usually use μMACS and MultiMACS GFP Isolation Kits (Miltenyi Biotec), consisting of anti-GFP antibody beads, M Column and μMACS separator, because of low background binding [5, 8]. Self-made anti-GFP affinity columns can be also used.
4. We usually use GL-Tip SDB and GL-tip GC columns (GL Sciences) for peptide desalting because of their high retentivity and recovery [5, 8]. Any other desalting columns can be used.
5. If necessary, the plants can be treated with 0.8 M mannitol or 50 μM ABA for 10 min to 5 h at 22 °C prior to grinding.
6. If necessary, the immunoprecipitated is checked by immunoblotting or silver staining as previously described [5, 8].
7. Ten microlitres of liquid sample is concentrated using 0.1% (v/v) formic acid solution at a flow rate of 1 μL/min over 10 min for sample concentration and fixation on the concentration column.
8. Elution and separation of the sample are performed at a flow rate of 500 nL/min with the following mobile phases: solvent A [2% (v/v) acetonitrile, 0.1% (v/v) formic acid] and solvent B [80% (v/v) acetonitrile, 0.1% (v/v) formic acid]. A linear gradient from A:B = 98:2 at 0 min to A:B = 60:40 in 300 min, A:B = 10:90 in 20 min, and A:B = 98:2 in 40 min is used in this analysis.
9. The ion spray voltage is set to 2300 V. A time-of-flight mass spectrometer (TOF-MS) is used to detect the ionized peptide fragments with a mass range of 400–1250 m/z over a signal accumulation period of 250 ms. In MS/MS analyses, high-

sensitivity mode is used with a collision energy of 37 V to detect the precursor ions with a mass range of 100–1600 m/z over a signal accumulation period of 100 ms per precursor ion. The MS/MS spectra of 20 types of precursor ions are acquired during the accumulation period. In addition, after the MS/MS spectrum of a precursor ion is acquired once, the ion is not reanalysed for 10 s.

10. The Paragon algorithm software (e.g., SCIEX ProteinPilot) is set up for protein identification with alkylation, reduction with iodoacetamide and trypsin digestion. The obtained results are statistically analysed using the false discovery rate (FDR) with the threshold $q < 0.05$.

References

1. Umezawa T, Nakashima K, Miyakawa T, Kuromori T, Tanokura M, Shinozaki K, Yamaguchi-Shinozaki K (2010) Molecular basis of the core regulatory network in ABA responses: sensing, signaling and transport. Plant Cell Physiol 51:1821–1839
2. Yoshida T, Christmann A, Yamaguchi-Shinozaki K, Grill E, Fernie AR (2019) Revisiting the basal role of ABA - roles outside of stress. Trends Plant Sci 24:625–635
3. Yoshida T, Mogami J, Yamaguchi-Shinozaki K (2014) ABA-dependent and ABA-independent signaling in response to osmotic stress in plants. Curr Opin Plant Biol 21:133–139
4. Nishimura N, Sarkeshik A, Nito K, Park SY, Wang A, Carvalho PC, Lee S, Caddell DF, Cutler SR, Chory J, Yates JR, Schroeder JI (2010) PYR/PYL/RCAR family members are major in-vivo ABI1 protein phosphatase 2C-interacting proteins in Arabidopsis. Plant J 61:290–299
5. Soma F, Mogami J, Yoshida T, Abekura M, Takahashi F, Kidokoro S, Mizoi J, Shinozaki K, Yamaguchi-Shinozaki K (2017) ABA-unresponsive SnRK2 protein kinases regulate mRNA decay under osmotic stress in plants. Nat Plants 3:16204
6. Soma F, Takahashi F, Suzuki T, Shinozaki K, Yamaguchi-Shinozaki K (2020) Plant Raf-like kinases regulate the mRNA population upstream of ABA-unresponsive SnRK2 kinases under drought stress. Nat Commun 11:1373
7. Shevchenko A, Tomas H, Havlis J, Olsen JV, Mann M (2006) In-gel digestion for mass spectrometric characterization of proteins and proteomes. Nat Protoc 1:2856–2860
8. Mogami J, Fujita Y, Yoshida T, Tsukiori Y, Nakagami H, Nomura Y, Fujiwara T, Nishida S, Yanagisawa S, Ishida T, Takahashi F, Morimoto K, Kidokoro S, Mizoi J, Shinozaki K, Yamaguchi-Shinozaki K (2015) Two distinct families of protein kinases are required for plant growth under high external Mg2+ concentrations in Arabidopsis. Plant Physiol 167: 1039–1057

Chapter 15

SWATH-MS Proteomic Approach to Discover Novel Protein Targets and Pathways in Response to Abscisic Acid

Bei Gao, Mo-Xian Chen, and Fu-Yuan Zhu

Abstract

SWATH-MS proteomic approaches enable the identification and quantification of thousands of proteins within a single profiling experiment, which is useful for the identification of genes regulated by abscisic acid (ABA) in a high-throughput manner. Here we describe the experimental procedures for protein extraction, digestion, peptides desalting, followed by the establishment of a DDA spectrum database and DIA-based SWATH detection and protein quantification. This method is able to identify and quantify proteins involved in ABA metabolism, signal perception and transduction with high accuracy and reproducibility.

Key words Abscisic acid, *Arabidopsis thaliana*, DDA, DIA, Label-free, Proteome, Rice

1 Introduction

Proteomics bridges genomics/transcriptomics and metabolomics, mass spectrometry (MS)-based proteomics have become an indispensable toolkit to investigate gene functions [1]. Peptide labeling coupled with data dependent acquisition (DDA) pipeline was prevalent in proteome quantification studies (e.g., [2]), which suffers restricted sample sizes because of limited types of isobaric tags and lower reproducibility due to batch effects and the selective nature of DDA. SWATH-MS (sequential window acquisition of all theoretical fragment ion spectra mass spectrometry), a proteomic procedure employing label-free coupled with data independent acquisition (DIA), was previously widely utilized in animal systems for years [3]. For plant tissues, the sodium deoxycholate method has been demonstrated to be a reliable approach for protein extraction followed by SWATH-MS detection [4], which was then successfully utilized in plant proteomic studies such as in *Arabidopsis thaliana* [5, 6] and rice [7]. SWATH-MS was also postulated to be a promising backbone technique for other plant molecular research

Takuya Yoshida (ed.), *Abscisic Acid: Methods and Protocols*,
Methods in Molecular Biology, vol. 2462, https://doi.org/10.1007/978-1-0716-2156-1_15,

areas besides quantitative proteomics [8, 9], while challenges still exist to achieve these [10].

Abscisic acid (ABA) is an important plant hormone that can mediate a variety of developmental and stress responses in plants. The accumulation of ABA was reported to be related to a number of important agronomic traits, such as apple fruit growth [11] and anthocyanin accumulation [12], while it remains unclear whether ABA is associated with alteration of protein profiles, such as apple fruit skin proteome data obtained by a SWATH-MS analysis [13]. By contrast, ABA treatment has been demonstrated to alter transcript alternative splicing events in *Arabidopsis* [14]. In this scenario, investigating the proteomic responses to ABA quantitively will be of great value to probe potential novel targets in future studies. For label-free proteomic quantifications, the normalized spectral abundance factor (NSAF) approach using spectral counting was proposed by taking account of protein length and library size [15], which was updated to distributed NSAF (dNSAF) approach by assigning multialigned peptides/spectral based on the amount of unique peptides/spectral counts for each protein [16]. In this way, dNSAF provided a practical approach to handle proteins encoded by duplicated homologous genes and especially alternatively spliced transcript isoforms. Thus, we summarized the experimental and data analytical procedures for SWATH-MS-based protein quantification, to facilitate future quantitative proteomic studies as well as associated biological processes in plants.

2 Materials

All solutions are prepared with ultrapure water and analytical grade reagents, and can be stored at room temperature without special instructions.

2.1 Protein Extraction

1. Tissue lysate solution: 2.5% SDS, 100 mM Tris-HCl, pH 8.0. Add 100 mL of water to a 1 L beaker, weigh 12.114 g Tris base and 25 g SDS into it, and add water to 900 mL. After mixing, adjust the pH to 8.0 with HCl, dilute to 1 L volume, and store at room temperature.
2. Acetone (*see* **Note 1**).
3. 1 M Tris-HCl buffer stock solution, pH 8.0.
4. Protein lysis buffer: 8 M urea, 100 mM Tris-HCl. Add 5 mL of 1 M Tris-HCl buffer to a 50 mL beaker, weigh 24 g of urea and add to it, make up to 45 mL by water, mix until Urea is completely dissolved, and dilute to 50 mL.
5. Protein assay reagent.
6. Freshly prepared or frozen plant tissue samples: Stored in liquid nitrogen before the following experimental procedures.

2.2 Enzymatic Digestion in Solution

1. DTT stock solution: 200 mM aqueous solution, stored at −20 °C.
2. Iodoacetamide stock solution: 400 mM aqueous solution, protected from light at room temperature (*see* **Note 2**).
3. 100 mM Tris-HCl buffer: Diluted 10 times with 1 M Tris-HCl buffer.
4. 1 μg/μL Trypsin: Stored at −20 °C.

2.3 Desalting of Peptides and Mass Detection

1. 10% Trifluoroacetic acid (TFA) stock solution.
2. Acetonitrile (ACN).
3. Balance buffer: 0.1% TFA aqueous solution.
4. Elution buffer: 80% ACN, 0.1% TFA.
5. Loading buffer: 2% ACN, 0.1% formic acid.
6. Solvent A: 20 mM ammonium acetate in water.
7. Solvent B: 20 mM ammonium acetate in ACN.
8. Mobile phase A: 3% DMSO, 97% H_2O, 0.1% formic acid.
9. Mobile phase B: 3% DMSO, 97% ACN, 0.1% formic acid.

2.4 Apparatus and Equipment

1. Mortar.
2. 15 mL centrifuge tube.
3. Handheld homogenizer.
4. Water bath.
5. Ultrasonic cell disruptor.
6. EP tube.
7. Absorbent paper.
8. pH indicator strips.
9. Sep-Pak C18 desalting column.
10. StageTip C18 desalting column.
11. C18 column (5 μm, 4.6 mm × 150 mm).
12. Trap column (5 μm, 300 Å, 300 μm × 5 mm).
13. Analytic column (3 μm, 120 Å, 300 μm × 15 mm).
14. Vacuum centrifugal concentrator.
15. High-performance liquid chromatography (HPLC) (e.g., Agilent 1260 or equivalent).
16. Mass spectrometer sample plate.
17. Mass spectrometry spotting plate.
18. micoLC system (e.g., Eksigent microLC 415 or equivalent).
19. TripleTOF mass spectrometer (e.g., Sciex TripleTOF 5600$^+$ or equivalent).

2.5 Data Analysis Tool

1. MSFragger.
2. DIA-NN v1.7.12.

3 Methods

3.1 Protein Extraction

1. Take an appropriate amount of plant tissue sample (100–500 mg) in a precooled mortar and grind it with liquid nitrogen to a powder without obvious particles/fibers.
2. Transfer the tissue powder to a 15 mL centrifuge tube with the addition of tissue lysate solution according to the weight–volume ratio of 1:10, and use a handheld homogenizer for tissue disruption for 5 min.
3. Put the tubes in boiling water bath for 15 min.
4. Use an ultrasonic cell disruptor to sonicate at 20–30% power for 10 min.
5. Centrifuge at 12,000 × *g* for 15 min, and transfer the supernatant to a new EP tube.
6. Add 4 times volume of acetone precooled at −20 °C, and incubate it overnight at 4 °C.
7. Take out the sample, centrifuge at 12,000 × *g* for 15 min at 4 °C, discard the supernatant.
8. Add appropriate amount of cold acetone (1–3 mL), shake to disperse the protein precipitate completely, centrifuge at 12,000 × *g* for 15 min at 4 °C, discard the supernatant, and repeat this step for 3 times.
9. Lay the protein precipitation tube flat on absorbent paper in a fume hood, air dry the pellet (~5–10 min), and store at −20 °C for subsequent experiment.

3.2 Enzymatic Digestion in Solution

1. Protein lysate solution dissolves the protein pellet, keep the protein concentration at 0.5–2 mg/mL. Protein concentrations are determined by the Bradford method.
2. Add DTT stock solution to a final concentration of 10 mM DTT, and incubate with shaking at 37 °C for 1 h.
3. Add iodoacetamide stock solution to the final concentration of about 40 mM, and incubate with shaking for 30 min at room temperature (25 °C) in the dark.
4. Centrifuge at 12,000 × *g* for 5 min. If there is precipitation, transfer the supernatant to a new EP tube.
5. Add 4× volume of 100 mM Tris-HCl solution until the final concentration of Urea is less than 2 M, and mix well.
6. Add trypsin (trypsin/protein: 1:50 w/w), incubate overnight at 37 °C.

3.3 Peptide Desalting

1. Add 10% TFA about 1/23 of the solution volume, adjust the pH of the solution to 6 by pH indicator strips and centrifuge at 12,000 × *g* for 5 min (*see* **Note 3**).
2. According to the protein content in the solution, select a Sep-Pak C18 desalting column of suitable capacity, and the remaining samples are drained and stored at −80 °C for later use.
3. Add 1× column volume of ACN to the desalting column to make it naturally stay under the action of gravity (*see* **Note 4**).
4. Add 2× column volume of 0.1% TFA equilibrium desalting column by 2 times.
5. Add peptide solution.
6. Clean the desalting column with 2× column volume of 0.1% TFA solution.
7. Add about 1.3× column volume of the elution buffer to elute the peptide, and collect the eluent to a new 1.5 mL EP tube.
8. Take a small amount of each sample and mix it to form a mixed testing sample (*see* **Note 5**).
9. Use a vacuum centrifugal concentrator to drain all the samples, and store at −80 °C for later use.

3.4 DDA Spectrum Database Establishment

1. Divide the mixed sample into 60 fractions by HPLC (*see* **Note 6**), and combine them into 15 fractions at equal intervals (i.e., the 1, 16, 31, 46 components are combined into one, the 2, 17, 32, 47 components are combined into one, etc.) (Fig. 1). Use a vacuum centrifugal concentrator to drain the samples.
2. Use the StageTip C18 desalting column to desalt the final 15 fractions for subsequent HPLC separation, and use a vacuum centrifugal concentrator to drain the sample before use.
3. Dissolve the peptide sample obtained in the previous step with 12 μL of loading buffer, shake well for 10 min, centrifuge at 12,000 × *g* for 10 min, take 10 μL of the supernatant and transfer it to the mass spectrometer sample plate. The sample is then injected into the TripleTOF LC/MS system for detection and 5 μL is injected for each detection in the Trap-Elute mode (*see* **Note 7**). Pooled sample is run using a DDA mode wherein MS spectra are acquired across a range of 350–1500 *m/z* followed by the acquisition of MS/MS spectra of the top 40 most intense precursor ions with a charge state of $z = 2$–5.
4. Use MSFragger to process the raw data collected by mass spectrometry, select the default workflow, set the precursor mass tolerance to −20 to +20 ppm, and use the spectraST method to generate the spectrum database (*see* **Note 8**).

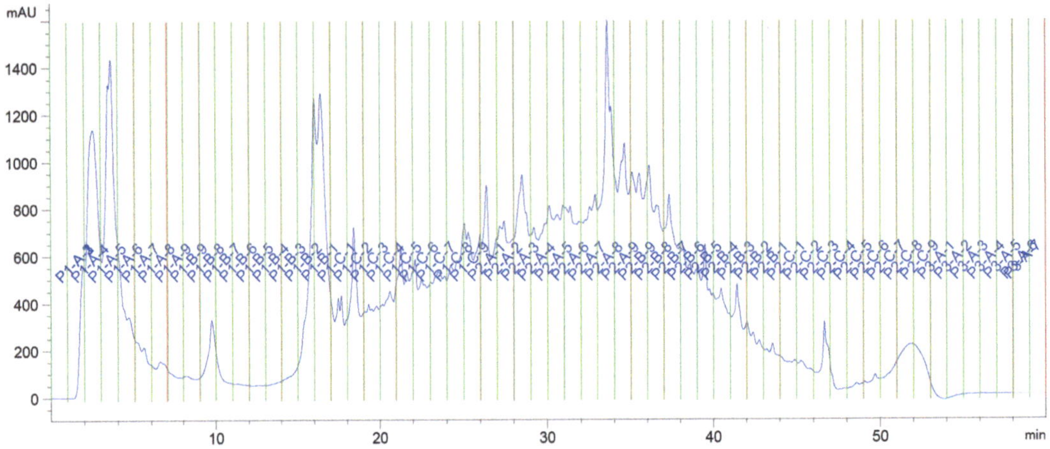

Fig. 1 The mixed samples were separated by high-pH reversed-phase chromatography with a 60 min gradient, and each fraction was collected every minute. The fractions separated by 15 min are combined with each other, and finally combined to obtain final separated 15 fractions

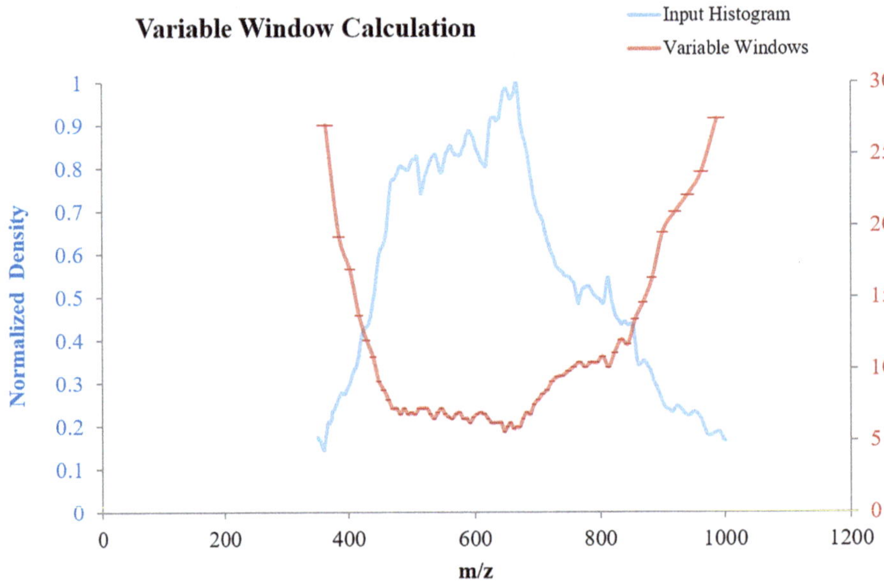

Fig. 2 Use the SWATH Variable Window Calculator toolkit to plan the variable window based on the TIC signal intensity distribution. After inputting the *m/z* distribution range, signal intensity distribution, number of windows and other parameters, the variable window distribution is calculated. The higher the signal intensity in the TIC (Total ion chromatogram), the denser the variable window distribution, and the narrower the *m/z* range contained in a single variable window

5. Extract the total ion chromatogram (TIC) generated by the mixed sample detection, and plan the variable window of the SWATH detection according to the ion signal intensity distribution (Fig. 2). Adjacent window detection crosses 1 *m/z* (*see* **Note 9**).

3.5 SWATH Detection and Protein Quantification

1. Dissolve the peptide sample to be tested with an appropriate amount of loading buffer to a concentration of about 1 μg/μL, shake well for 10 min, centrifuge at 12,000 × *g* for 10 min.
2. Transfer the supernatant to the mass spectrometry spotting plate, and inject the sample into the TripleTOF LC/MS system for detection with 5 μL injection volume for each analysis.
3. The liquid phase system is set with a 60 min gradient, and the liquid flow rate is 5 μL/min (*see* **Note 7**).
4. The mass spectrometer collects data in DIA mode, and the window setting is imported into the variable window planned in Subheading 3.4. The MS1 ion accumulation time is set to 250 ms, and the MS2 ion accumulation time is set to 30 ms.
5. Use DIA-NN v1.7.12 to extract SWATH quantitative information from the raw data collected by mass spectrometry. DIA-NN can set the parameters according to the default configuration, import the spectrum database generated in Subheading 3.4 and the proteomic sequence database of the species to which the sample belongs, select 2 for miss cleavages, and check the quantities matric. The test results are filtered with 1% false discovery rate (FDR). The final protein quantitative information is used for difference comparison and *t*-test analysis after $\log_2$ conversion, data filling, and data normalization procedures (Fig. 3) (*see* **Notes 10** and **11**).

4 Notes

1. Precool at −20 °C before use.
2. This solution needs to be freshly prepared.
3. At this time, the final concentration of TFA is about 0.43%.
4. In order to obtain a better desalting effect, the amount of solution added to the desalting column each time does not exceed 1 column volume, and then adding the next step or the remaining solution after completely entering the column, and it is necessary to ensure that the desalting column will not dry up in the process.
5. The total amount of mixed samples is not less than 200 μg.
6. Samples are loaded on a C18 column (5 μm, 4.6 mm × 150 mm) with 95% solvent A and 5% solvent B and separated in a gradient from 5% to 95% of solvent B for 45 min followed by 95% solvent A for 15 min at 0.2 mL/min.
7. Prepared peptides are loaded onto trap column (5 μm, 300 Å, 300 μm × 5 mm) followed by separation on a analytic column (3 μm, 120 Å, 300 μm × 15 mm) using a 60 min gradient at a flow rate of 5 μL/min. The detail LC conditions are as

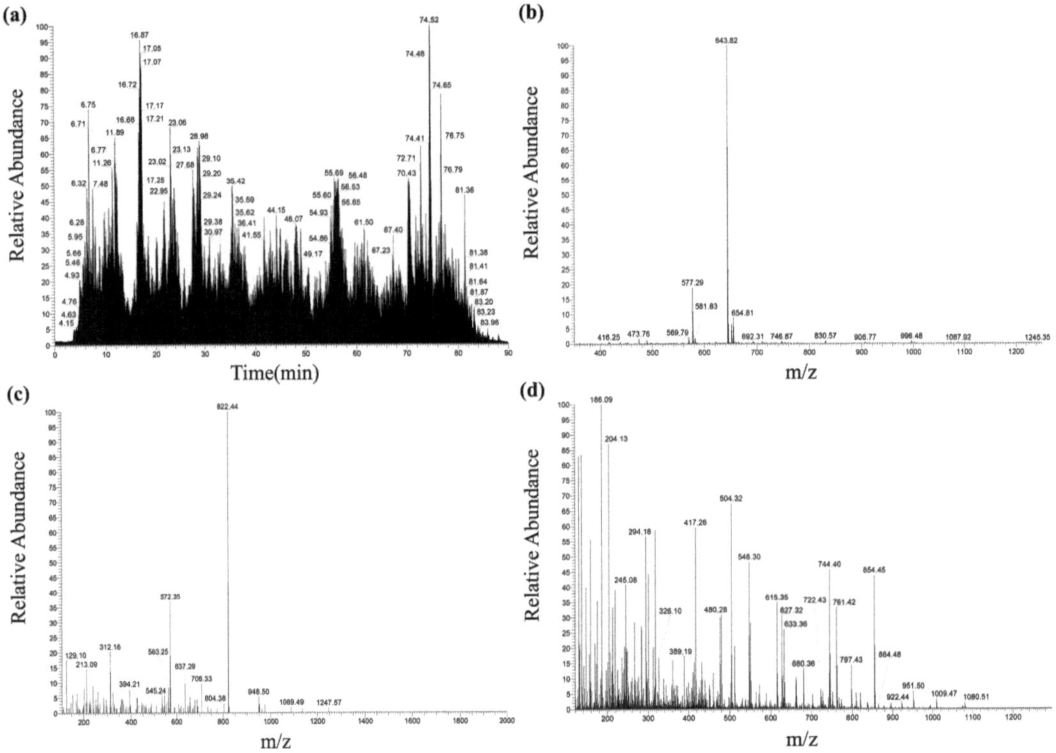

Fig. 3 MS spectrum information for protein quantification. (**a**) TIC chart (**b**) MS1 spectrum (**c**) MS2 spectrum by DDA scan (**d**) MS2 spectrum by DIA scan. The MS2 spectrum of the DDA scan is derived from the MS2 scan performed by the precursor ion with top signal intensity in the MS1 scan. The MS2 spectrum of the DIA scan is obtained according to a variable window, and all the precursor ions within specific ranges of mass-nucleus ratio (*m/z*) will undergo an MS2 scan

followings: 0 min in 5% mobile phase B and 95% mobile phase A, 35 min of 5–24% B, 10 min of 24–35% B, 6 min of 35–80% B, 80% B for 3 min, 1 min of 80–5% B, and a final step in 5% B for 5 min.

8. The file name of the generated spectrum database is "output_-file_irt_con.splib".
9. The number of variable windows needs to be set according to TIC peak width T, MS1 ion accumulation time TMS and MS2 ion accumulation time TMSMS, generally follow the variable window number N<(T/10-TMS)/TMSMS.
10. The suffix of the spectrogram database file generated in Subheading 3.4 needs to be changed to ".sptxt" to be recognized by DIA-NN.
11. For plants with a reference genome, a comprehensive proteome comprising isoforms is recommended, if not, protein sequences inferred from de novo transcriptomes (e.g., Trinity assembly and TransDecoder coding sequences) may also work.

Acknowledgments

This work was supported by the Natural Science Foundation of Jiangsu Province (SBK2020042924), Jiangsu Agricultural Science and Technology Innovation Fund (CX (21) 2023), Major project of natural science research in colleges of Jiangsu Province (20KJA220001), Platform Funding for Guangdong Provincial Enterprise Key Laboratory of Seed and Seedling Health Management Technology (2021B1212050011), the National Natural Science Foundation of China (32001452).

References

1. Gillet LC, Navarro P, Tate S, Rost H, Selevsek N, Reiter L, Bonner R, Aebersold R (2012) Targeted data extraction of the MS/MS spectra generated by data-independent acquisition: a new concept for consistent and accurate proteome analysis. Mol Cell Proteomics 11(6):O111.016717. https://doi.org/10.1074/mcp.O111.016717
2. Unwin RD, Griffiths JR, Whetton AD (2010) Simultaneous analysis of relative protein expression levels across multiple samples using iTRAQ isobaric tags with 2D nano LC-MS/MS. Nat Protoc 5(9):1574–1582. https://doi.org/10.1038/nprot.2010.123
3. Ludwig C, Gillet L, Rosenberger G, Amon S, Collins BC, Aebersold R (2018) Data-independent acquisition-based SWATH-MS for quantitative proteomics: a tutorial. Mol Syst Biol 14(8):e8126. https://doi.org/10.15252/msb.20178126
4. Scheerlinck E, Dhaenens M, Van Soom A, Peelman L, De Sutter P, Van Steendam K, Deforce D (2015) Minimizing technical variation during sample preparation prior to label-free quantitative mass spectrometry. Anal Biochem 490:14–19. https://doi.org/10.1016/j.ab.2015.08.018
5. Zhu FY, Chen MX, Chan WL, Yang F, Tian Y, Song T, Xie LJ, Zhou Y, Xiao S, Zhang J, Lo C (2018) SWATH-MS quantitative proteomic investigation of nitrogen starvation in Arabidopsis reveals new aspects of plant nitrogen stress responses. J Proteome 187:161–170. https://doi.org/10.1016/j.jprot.2018.07.014
6. Zhu FY, Chan WL, Chen MX, Kong RP, Cai C, Wang Q, Zhang JH, Lo C (2016) SWATH-MS quantitative proteomic investigation reveals a role of jasmonic acid during lead response in arabidopsis. J Proteome Res 15(10):3528–3539. https://doi.org/10.1021/acs.jproteome.6b00258
7. Chen MX, Zhu FY, Gao B, Ma KL, Zhang Y, Fernie AR, Chen X, Dai L, Ye NH, Zhang X, Tian Y, Zhang D, Xiao S, Zhang J, Liu YG (2020) Full-length transcript-based proteogenomics of rice improves its genome and proteome annotation. Plant Physiol 182(3):1510–1526. https://doi.org/10.1104/pp.19.00430
8. Chen MX, Zhang Y, Fernie AR, Liu YG, Zhu FY (2021) SWATH-MS-based proteomics: strategies and applications in plants. Trends Biotechnol 39(5):433–437. https://doi.org/10.1016/j.tibtech.2020.09.002
9. Chen MX, Mei LC, Wang F, Boyagane Dewayalage IKW, Yang JF, Dai L, Yang GF, Gao B, Cheng CL, Liu YG, Zhang J, Hao GF (2020) PlantSPEAD: a web resource towards comparatively analysing stress-responsive expression of splicing-related proteins in plant. Plant Biotechnol J. https://doi.org/10.1111/pbi.13486
10. Zhu FY, Song YC, Zhang KL, Chen X, Chen MX (2020) Quantifying plant dynamic proteomes by SWATH-based mass spectrometry. Trends Plant Sci 25(11):1171–1172. https://doi.org/10.1016/j.tplants.2020.07.014
11. Berüter J (1983) Effect of abscisic acid on sorbitol uptake in growing apple fruits. J Exp Bot 34(6):737–743. https://doi.org/10.1093/jxb/34.6.737
12. An JP, Yao JF, Xu RR, You CX, Wang XF, Hao YJ (2018) Apple bZIP transcription factor MdbZIP44 regulates abscisic acid-promoted anthocyanin accumulation. Plant Cell Environ 41(11):2678–2692. https://doi.org/10.1111/pce.13393
13. Chen M-X, Sun C, Zhang K-L, Song Y-C, Tian Y, Chen X, Liu Y-G, Ye N-H, Zhang J, Qu S, Zhu F-Y (2019) SWATH-MS-facilitated proteomic profiling of fruit skin between Fuji apple and a red skin bud sport mutant. BMC

Plant Biol 19(1):445. https://doi.org/10.1186/s12870-019-2018-1

14. Zhu FY, Chen MX, Ye NH, Shi L, Ma KL, Yang JF, Cao YY, Zhang YJ, Yoshida T, Fernie AR, Fan GY, Wen B, Zhou R, Liu TY, Fan T, Gao B, Zhang D, Hao GF, Xiao S, Liu YG, Zhang JH (2017) Proteogenomic analysis reveals alternative splicing and translation as part of the abscisic acid response in Arabidopsis seedlings. Plant J 91(3):518–533. https://doi.org/10.1111/tpj.13571
15. Zybailov B, Mosley AL, Sardiu ME, Coleman MK, Florens L, Washburn MP (2006) Statistical analysis of membrane proteome expression changes in Saccharomyces cerevisiae. J Proteome Res 5(9):2339–2347. https://doi.org/10.1021/pr060161n
16. Zhang Y, Wen ZH, Washburn MP, Florens L (2010) Refinements to label free proteome quantitation: how to deal with peptides shared by multiple proteins. Anal Chem 82(6):2272–2281. https://doi.org/10.1021/ac9023999

Chapter 16

Genome-Wide Association Analyses to Identify SNPs Related to Drought Tolerance

Shengxue Liu and Feng Qin

Abstract

Drought stress is a serious agronomic problem resulting in significant yield losses globally. Breeding cultivars with drought tolerance is an important strategy that can be used to address this problem. Drought tolerance, however, is a complex multigenic trait, making advancements with conventional breeding approaches very challenging. This emphasizes the importance of dissecting the genetics of this trait and the identification and cloning of genes responsible for drought tolerance. With the rapid development of sequencing technologies and analytic methodologies, genome-wide association study (GWAS) has become an important tool for detecting natural variations underlying complex traits in crops. Identified loci can serve as targets for genomic selection or precise editing that enables the molecular design of new cultivars. This chapter describes the pipeline of statistical methods used in GWAS analysis, and covers field design, quality control, population structure control, association tests, and visualization of data. GWAS methodology used to dissect the genetic basis of drought tolerance is presented, and perspectives for optimizing the design and analysis of GWAS are discussed. The provided information serves as a valuable resource for researchers interested in GWAS technology.

Key words GWAS, Drought tolerance, Natural variation, Crops, Mixed model, Genotyping

1 Introduction

Drought is a frequent climate event characterized by below-normal precipitation over a period of months to years. Plants alter their physiology to adapt to moisture gradients in soil, modify root growth and architecture, and close stomata on their aboveground organs [1]. Tissue-specific responses to drought include stunted growth and reduced yield. Drought resistance can be defined as the ability of a plant to sense a water deficiency and subsequently signal and initiate adaptive strategies at the whole plant level. It is a complex trait that encompasses three mechanisms: "drought escape" (acceleration of the plant reproductive phase or undergoing a period of dormancy until suitable conditions return), "drought avoidance" (the ability to maintain a relatively high tissue

Takuya Yoshida (ed.), *Abscisic Acid: Methods and Protocols*,
Methods in Molecular Biology, vol. 2462, https://doi.org/10.1007/978-1-0716-2156-1_16,

water content to prevent tissue damage), and "drought tolerance" (ability to maintain cellular homeostasis through adaptive traits over the course of the drought period) [2]. With respect to drought tolerance at a cellular level, drought stress-related signal transduction networks involving abscisic acid (ABA) dependent and independent pathways are activated and transduced after stress signals are perceived by plant cells [3]. Numerous genes function in stress response and stress tolerance involves alterations at the molecular, cellular, and physiological levels. Plants exhibit numerous physiological and morphological changes that facilitate their ability to adapt to drought stress, including stomatal closure, leaf rolling, and enhanced or altered root growth [4]. Therefore, elucidating the mechanisms which allow a plant to sustain its during episodes of drought stress can provide strategies to improve stress tolerance through the genetic improvement of crops.

Linkage mapping and genome-wide association study (GWAS) are two major strategies that can be used to unravel the genetic basis of complex traits in plants, and identify quality trait loci (QTLs) associated with the trait in question [5]. The basis for linkage mapping is genetic recombination. There are two major limitations, however, to linkage mapping in conventional recombinant populations. First, there is normally an insufficient number of recombination events in a mapping population comprising hundreds of descendants, which results in poor mapping resolution. Second, only QTLs which contain the sequence divergence between the selected parents can be detected. In contrast, GWAS utilizes historical and evolutionary recombinations to identify the genetic loci underlying traits, and provides a relatively high-resolution mapping of a trait [6]. With the rapid development of sequencing technologies and analytic methodology, GWAS has been conducted in many plant species, including maize, rice, sorghum, and foxtail millet [7–12], although over the past decade, rice and maize are the major crops that have utilized GWAS. In rice, the genome-wide linkage disequilibrium (LD) decay rates were estimated to be ~123 kb and ~167 kb, at r^2 of 0.25 and 0.28, respectively [13]. In contrast, maize, being an outcrossing species, exhibits a rapid LD decay (within ~5.5 kb, $r^2 < 0.2$) [14]. The GWAS resolution in maize is higher than it is in rice, and may reach the single-gene level. This makes maize an ideal crop for GWAS and significant progress has been achieved. GWAS has been implemented to dissect drought tolerance in maize and to facilitate the cloning of causative genes [10].

The significant reduction in the cost of sequencing, and the development of new omics technologies, such as transcriptomics, metabolomics, and proteomics, has resulted in an exponential growth in the amount available-omics data. Thus, there is an important need to integrate multiomic data into association mapping to bridge the gap between genotypic and phenotypic variation. Omics-based technology has been used to identify candidate genes, thereby assisting in QTL mapping [15, 16]. The combined analysis of multi-omics data will foster a change from GWAS to omic-wide association studies. Such an approach will provide an integration of omics data and will also effectively deepen our understanding of the balance between stress response and grain yield and quality. Innovative methodologies and statistical methods will accelerate the molecular design of new cultivars that will minimize the risk of global food insecurity. This chapter introduces GWAS methodology for identifying drought-tolerant loci, and covers field design, quality control, population structure control, association tests, and data visualization.

2 Materials

2.1 Software Packages

1. Standard personal computer that can run R version 3.5.3 or above (https://www.r-project.org/).
2. The Linux systems that that can run Trimmomatic, FASTX-Toolkit, BWA-MEM, GATK, EINGENSTRAT, etc. (https://www.linux.org/).
3. Software for filtering RNA-seq reads: Trimmomatic or FASTX-Toolkit (*see* **Note 1**).
4. Software for mapping: BWA-MEM or HISAT2 ver.2.0.5 (*see* **Note 2**).
5. Software for SNPs detection: Genome Analysis Toolkit (GATK).
6. Software for population structure estimation: EINGENSTRAT (https://www.hsph.harvard.edu/alkes-price/software/) (*see* **Note 3**).
7. Software for population structure estimation: STRUCTURE (https://web.stanford.edu/group/pritchardlab/structure_software/release_versions/v2.3.4/html/structure.html) (*see* **Note 3**).
8. SPAGeDi1-5a or above (https://ebe.ulb.ac.be/ebe/SPAGeDi.html) (*see* **Note 4**).
9. Excel, a spreadsheet software program for data visualization and analysis tool (https://support.microsoft.com/).
10. RStudio, a set of integrated tools designed to help you be more productive with R (https://rstudio.com/).

2.2 Transcriptomics and Phenotype Data

1. The sequencing data of immature seeds at 15 days after pollination from 368 maize inbred lines (*see* **Note 5**).
2. The phenotype data of maize seedlings (*see* **Note 6**).

3. The maize genome sequence: B73_RefGen_v4 (ftp://ftp.ensemblgenomes.org/pub/plants/release-38/fasta/zea_mays/dna/).

3 Methods

3.1 Study Design for GWAS

An appropriately designed study will avoid systematic bias in the subsequent analysis and improve the statistical power available to detect association signals. Presently, five basic elements in a GWAS design are discussed, namely, choosing a suitable trait and assembling a mapping population, field design, genotyping for association mapping, statistical analysis, and lastly, phylogenetic and population genetic analyses.

3.1.1 Identifying a Suitable Target Trait and Assembling a Mapping Population

1. Identify the target trait that have a wide range of diversity in the assembled population considering the followings: (1) Usually genotypes or germplasms from different resources or origins need to be included to ensure a sufficient level of diversity. (2) It should also be possible to accurately and repeatedly quantify the target trait of interest with a high degree of reproducibility. (3) Background knowledge on physiological and/or developmental effects on the target trait is desirable, although not essential. Specifically regarding drought resistance, several component traits have been identified and analyzed (Table 1) (*see* **Note 7**).
2. Assemble an appropriate panel for GWAS, which harnesses the genetic diversity present in natural populations to potentially resolve complex trait variation to single genes or SNPs (*see* **Note 8**).

3.1.2 Experimental Design for Phenotyping

1. Design experiments for comparative tests of drought adaptation for seedlings (*see* **Note 9**), under field conditions (*see* **Note 10**) or by applying osmotic stress (*see* **Note 11**).

3.1.3 Obtaining High-Quality Genotype Data

1. Obtain a great number of molecular markers for the mapping of QTLs, especially for species with a quick LD decay, either by DNA hybridization or polymerase chain reaction (PCR) techniques (*see* **Note 12**) or high-throughput sequencing technology (*see* **Note 13**).

3.1.4 Population Analyses

1. Group individuals used in association mapping by population structure and within-group familial relatedness due to domestication and selection by breeders (*see* **Note 14**). In general, the population structure (*Q*) (*see* **Note 15**) and familial relatedness (*K*) (*see* **Note 16**) can be estimated by many software.

Table 1
Examples of association mapping studies in a variety of crop species

Trait category	Phenotype	Species	Sample size	Marker numbers	Candidate gene	References
Seedling stage	SR	Maize	368	525,105	*ZmDREB2.7*	Liu et al. [9]
		Maize	368	556,944	*ZmNAC111*	Mao et al. [17]
		Maize	367	556,944	*ZmVPP1*	Wang et al. [10]
		Maize	368	525,105	*ZmPP2C-A10*	Xiang et al. [18]
Components of grain yield	ASI	Maize	961	2,052	65 (GRMZM2G164325, etc.)	Lu et al. [19]
	GY, DTA, DTS, ASI, etc.	Maize	350	56,110	33 (GRMZM2G125777, etc.)	Xue et al. [20]
	GY, ASI, EG, EL, etc.	Maize	240	29,619	108 (GRMZM2G418217, etc.)	Thirunavukkarasu et al. [21]
	GY, DTA, DTS, ASI, etc.	Maize	346	60,000	16 (GRMZM2G035688, etc.)	Farfan et al. [22]
	GY	Rice	175	150,325	50 (LOC_Os01g43060, etc.)	Pantaliao et al. [23]
	GY, PH	Rice	270	1,019,883	9 (OsRLK5, etc.)	Ma et al. [24]
	GYP, PH, SL, TKW, etc.	Wheat	277	395,681	295 (TaWRKY8, etc.)	Li et al. [25]
	GYP, PH, DTH, DTM, etc.	Wheat	320	9,626	18 (TraesCS4B02G051500, etc.)	Gahlaut et al. [26]
	GY, PH, PL, HI, etc.	Rice	280	215,250	38 (LOC_Os02g30270, etc.)	Bhandari et al. [12]
	PH, GY, TN, SL, etc.	Wheat	290	15,737	None	Abou-Elwafa and Shehzad [27]
Root traits	TNR, RDW, CVA, PRL, etc.	Maize	384	681,257	GRMZM2G153722	Pace et al. [28]
	RDW, RL, THK, PDW, etc.	Rice	180	22,000	23 (Os01g05820, etc.)	Phung et al. [29]
	RDW, RL, RSA, RV, etc.	Maize	396	955,690	98 (GRMZM2G148106, etc.)	Zaidi et al. [30]

(continued)

Table 1 (continued)

Trait category	Phenotype	Species	Sample size	Marker numbers	Candidate gene	References
	RL, RV, RDW.	Maize	529	4,358,600	11 (Nal1, etc.)	Li et al. [11]
	RHL	Maize	367	556,944	*ZmTIP1*	Zhang et al. [31]
	PRL, SRL, SRN.	Maize	209	43,252	7 (GRMZM2G136364, etc.)	Guo et al. [32]
Omics	Gene expression, SR	Maize	224	1,288,889	97 (Zm00001d051554, etc.)	Liu et al. [16]

SR survival rate, *DTA* days to anthesis, *DTS* days to silking, *ASI* anthesis-silking interval, *GY* grain yield, *EG* ear girth, *EL* ear length, *PH* plant height, *GYP* grain yield per plant, *SL* main spike length, *TKW* thousand kernel weight, *DTH* days to heading, *DTM* days to maturity, *PL* panicle length, *HI* harvest index, *TN* tiller number, *TNR* total number of roots, *RDW* root dry weight, *CVA* convex root area, *PRL* primary root length, *RL* root length, *THK* root thickness, *PDW* plant dry weight, *RSA* root surface area, *RV* root volume, *RHL* root hair length, *SRL* seminal root length, *SRN* seminal root number

3.1.5 Statistical Models for GWAS

1. Choose appropriate model(s) for statistical analysis (*see* **Note 17**). In general, six models have been used to uncover the associations between the genetic variations associated with complex traits:
   (a) The General Linear Model (GLM) model, without controlling *Q* and *K* (*see* **Note 18**).
   (b) The *Q* model includes *Q* as a cofactor, to control the spurious associations (*see* **Note 19**).
   (c) The principal component analysis (PCA) model uses PCA for the *Q* estimation (*see* **Note 19**).
   (d) The *K* model includes *K* as a cofactor, to control false-positive associations (*see* **Note 20**).
   (e) The *Q* + *K* model controls both *Q* and *K* (*see* **Note 21**).
   (f) The PCA + *K* model controls both *Q* and *K*, through PCA and *K* (*see* **Note 21**).

3.2 A Case Study of Drought-Tolerant SNP Identification

An overview of the analytical steps and processes in GWAS used to detect genetic loci controlling drought resistance in maize seedlings can be provided for a panel of 368 maize inbred lines genotyped by deep RNA-seq [10, 33]. The overview includes the generation of genotyping data (RNA-sequencing, reads quality control, reads mapping, and SNP calling), control of population stratification, association analysis, and other related issues.

3.2.1 Quality Control of Sequencing Reads

1. Obtain FASTA or FASTQ files as data output, containing multiple short-reads sequences.
2. RNA-seq reads are filtered to exclude reads with low sequencing quality and adapter sequences using Trimmomatic or FASTX-Toolkit software. The command lines used in the two programs are as follows.
   (a) Running Trimmomatic for Paired End Mode.

```
java -jar trimmomatic-0.35.jar PE -phred33 input_forward.fq.gz
input_reverse.fq.gz output_forward_paired.fq.gz output_for-
ward_unpaired.fq.gz output_reverse_paired.fq.gz output_re-
verse_unpaired.fq.gz ILLUMINACLIP :TruSeq3-PE.fa:2:30:10
LEADING:3 TRAILING:3 SLIDINGWINDOW:4:15 MINLEN:36
```

   (b) Running FASTX for Paired End Mode.

```
fastq_quality_filter [-q N] [-p N] [-i INFILE] [-o OUTFILE]
```

3.2.2 Read Mapping

1. All of the high-quality reads are mapped to the B73_RefGen_v4 using BWA-MEM or HISAT2 ver.2.0.5 as follows.
2. BWA-MEM command for reads mapping.

```
bwa mem -M -t [threads] [ref] [R1_clean].fq [R2_clean].fq |
samtools view -bS > [name.source].bam
```

3. Quality control for alignment.

```
samtools view -h [name.source].bam | samtools view -bS -q30 >
[name].bam
```

4. Sorting alignment.

```
samtools sort [name].bam [name].sorted.bam
```

5. Index alignment.

```
samtools index [name].sorted.bam
```

6. Checking sequence depth and breadth of coverage.

```
bedtools genomecov -ibam [name].sorted.bam > [name].coverage
samtools depth -a [name].bam > [name].depth
```

7. Only reads that mapped uniquely to the genome are retained for further variation calling.
8. Alignment results are then sorted according to their alignment position on a chromosome and converted to a SAM (Sequence Alignment/Map) format.

3.2.3 SNP Calling

1. The alignment results of each germplasm are merged into one file in BAM format.
2. This file is then used to detect SNPs using the GATK (GATK HaplotypeCaller for SNP calling) as follows.
3. Creating GVCF files of each germplasm.

```
java -Xmx30g -XX:ParallelGCThreads=4 -Djava.io.tmpdir=./tmp
[i] -jar GenomeAnalysisTK.jar -T HaplotypeCaller -R [ref] -I
[name].sorted.bam -o [name].g.vcf -nct 4 --emitRefConfidence
GVCF
```

4. SNP detection from GVCF files.

```
java -Xmx30g -XX:ParallelGCThreads=4 -Djava.io.tmpdir=./tmp
[i] -jar GenomeAnalysisTK.jar -T GenotypeGVCFs -R [ref] -V
[name1].g.vcf -V [name2].g.vcf ... -nct 4 -o [out].vcf
```

5. SAMtools for SNP detection and filtering.

```
bcftools mpileup [name].sorted.bam -f [ref].fa | bcftools call
-cv -o [raw].vcf
bcftools view [raw].vcf | misc/vcfutils.pl varFilter > [name-
filtered].vcf
vcftools --vcf [vcf] [--plink] --max-missing 0.8 --maf 0.05
[--remove-indels] --out [outfile]
```

3.2.4 Control of Population Stratification

Various degrees of population stratification may exist despite careful study design and sample collection. An unknown or uncontrollable mixture of ethnicity groups would result in allele frequency discrepancies, inflated variances, and increased spurious associations.

1. Principal components (PC) of the association panel are calculated by EIGENSTRAT using the high-quality SNPs.
   (a) Data format conversion
   Using VCFtools convert VCF files to ped and map files. Hapmap file is converted directly into ped and map files through tassel (*see* **Note 22**).
   (b) Running smartpca
   The perl script calls the smartpca program, as follow:

```
#!/usr/bin/perl
#$ENV{'PATH'} = "../bin:$ENV{'PATH'}";
$command = "smartpca";
$command .= "-p par.example > example.log";
print("$command\n");
system("$command");
$command = "ploteig";
$command .= " -iexample.evec ";
$command .= " -c 1:2 ";
$command .= " -p Control ";
$command .= " -x ";
$command .= " -o example.plot.xtxt "; # must end in .xtxt
print("$command\n");
system("$command");
$command = "evec2pca.perl 10 example.evec  example.ind  ex-
ample.pca";
print("$command\n");
system("$command");
```

2. SPAGeDi for kinship coefficient

   Running SPAGeDi1-5a (build04-03-2015).exe with the following steps and parameters:

   (a) Specifying the data and results files.

       It is advised to create the data file using a worksheet program such as Excel and then save it as a "tab delimited text file". It is recommended to read the user's manual for details.

   (b) Specifying level of analysis and statistics.

       1-KINSHIP coefficient.

   (c) Specifying computational options.

       4-Jackknife over loci.

   (d) Specifying output options.

       4-report matrices with pairwise spatial distances and genetic coefficients.

   (e) Format for pairwise spatial and genetic distances.

       3-multilocus estimates only (matrix and columnar forms).

3.2.5 Association Tests and Other Related Issues

The Mixed Linear Model (MLM) model is undertaken for the association mapping taking into account the population structure and familiar relatedness [34] (*see* **Note 23**).

1. Running run_pipeline.pl to execute the pipeline in Linux for GWAS.

```
./run_pipeline.pl -Xms8g -Xmx8g -fork1 -h <hapmap file> -fork2
-r <phenotype file> -fork3 -q <population structure file>
-fork4 -k <kinship file> -combine5 -input1 -input2 -input3
-intersect -combine6 -input5 -input4 -mlm -mlmCompressionLevel
None -mlmVarCompEst <P3D (default) or EachMarker> -export
<filename> -runfork1 -runfork2 -runfork3 -runfork4
```

2. Running GWAS in the R software environment for GWAS.

```
library(MASS)
library(multtest)
library(gplots)
library(compiler)
library(scatterplot3d)
source("http://www.zzlab.net/GAPIT/emma.txt")
source("http://www.zzlab.net/GAPIT/gapit_functions.txt")
#Step 1: Set data directory and import files
setwd("<working directory> ")
myY <- read.table("<phenotype >", head = TRUE)
myG <- read.delim("<genotype data in hapmap format >", head =
FALSE)
```

```
myKI <- read.table("kinship matrix", head = FALSE)
myCV <- read.table("Q matrix", head = TRUE)
#Step 2: Run GAPIT
MyGAPIT <- GAPIT(Y=myY, G=myG, KI=myKI, CV=myCV)
```

3. LD decay analysis by plink.

```
plink --file [name] --r2 --ld-window 99999 --ld-window-r2 0 --
ld-window-kb 1000 --out [fileouts],
```

4. Manhattan plot for summarizing GWAS results in an R environment.

```
setwd("working directory ")
file = read.table(file="gwas_result", header=T, sep="\t")
traitname = "figure name"
png(file=paste(""figure name".png", sep=""), res=120,
width=800, height=800)
plot(c(0,2100000000), c(0,12), type="n", xlab="Chr",
     ylab=expression(-log[10]~~(P[observed])), xaxs="i", yax-
s="i", axes=F,
      mgp=c(3,1,0), bty="l", cex.lab=2, lwd=2, ylim=c(0,12),
mgp=c(2,1,0))
file = file[-1, ]
file = file[order(file[, 3]), ]
maxy = -log10(sort(file[,7])[1])+1
re = 0
k = "midnightblue"
for (j in 1:10){
subx = file[which(file[, 3]==j), ]
subx = subx[order(subx[, 4]), ]
subx[,4]=re+subx[, 4]
points(subx[, 4], -log10(subx[, 7]), col=k, xaxt="n", pch=16,
cex=1)
     if (k="midnightblue") {k="skyblue"} else { k="midnight-
blue"}
      border=c(307041717, 551483993, 787151827, 1034146432,
1258048672,
              1432081842, 1614463384, 1795586021, 1955355803,
2106338117)
    re =border[j] # the B73 reference genome RefGen_V4}
axis(1, c(153520858, 429262855, 669317910, 910649129,
1146097552, 1345065257,
     1523272613, 1705024703, 1875470912, 2030846960), 1:10,
lwd=2,
    cex.axis=1.9, cex.lab=1.9)
abline(v=0)
```

```
axis(2, c(0, 2, 4, 6, 8, 10, 12), c(0,2,4,6,8,10,12), cex.
axis=1.9, lwd = 2, las=1)
abline(h=5, lty=2, lwd=2, col="red")
dev.off()
```

5. QQ-plot to evaluate the model of GWAS in an R environment.

```
setwd("working directory")
file1 = read.table(file=" gwas_result", header=T, sep="\t")
filename = "gwas "
maxy1 = -log10(sort(file1[,7])[1])+1
PPX1 = sort(-log10(file1[,7]))
NY1 = -log10((1:length(PPX1))/length(PPX1))
PPY1 =sort(NY1)
png(file=paste("QQplot_", filename,".png",sep=""), res =140,
width=800,
height=800)
plot(PPY1, PPX1, xaxt="n", yaxt="n", xlab=expression(-log[10]
~~(P[predicted])),
     ylab=expression(-log[10]~~(P[observed])), xaxs="i", yax-
s="i", axes=F,
     col="darkblue", ylim=c(0, 12), xlim=c(0, 6), pch=16,
cex=0.8, cex.axis=1.2,
   cex.lab=1.2)
points(seq(0, 6, 0.00001), seq(0, 6, 0.00001), xaxt="n",
yaxt="n", xlab="", ylab="",
   xaxs="i", yaxs="i", axes=F, col="gray", ylim=c(0, 6), xlim=c
(0, 6), type="l",
   lwd=3, cex.axis=1.5, cex.lab=1.1)
par(new=T)
axis(2, c(seq(0, 12, by=2)), lwd=3, cex.axis=1.9, cex.lab=1.9)
axis(1, c(0, 1, 2, 3, 4, 5, 6), lwd=3, cex.axis=1.9, cex.
lab=1.9)
dev.off()
```

3.2.6 Data Presentation and Interpretation

Here we present an example result of the GWAS of maize drought tolerance at the seedling stage that identified 83 genetic variants at a significant threshold ($p < 1.0 \times 10^{-5}$), under the MLM model (Fig. 1).

1. Create the PCA plot showing all lines clustering into SS (stiff stock), NSS (nonstiff stock), and TST (tropical/subtropical) subpopulations, with the germplasms of mixed origin dispersing among these subpopulations (Fig. 1a).
2. Create the Manhattan plot for the GWAS, in which the SNPs located within *ZmVPP1* are labeled as red dots (Fig. 1b).

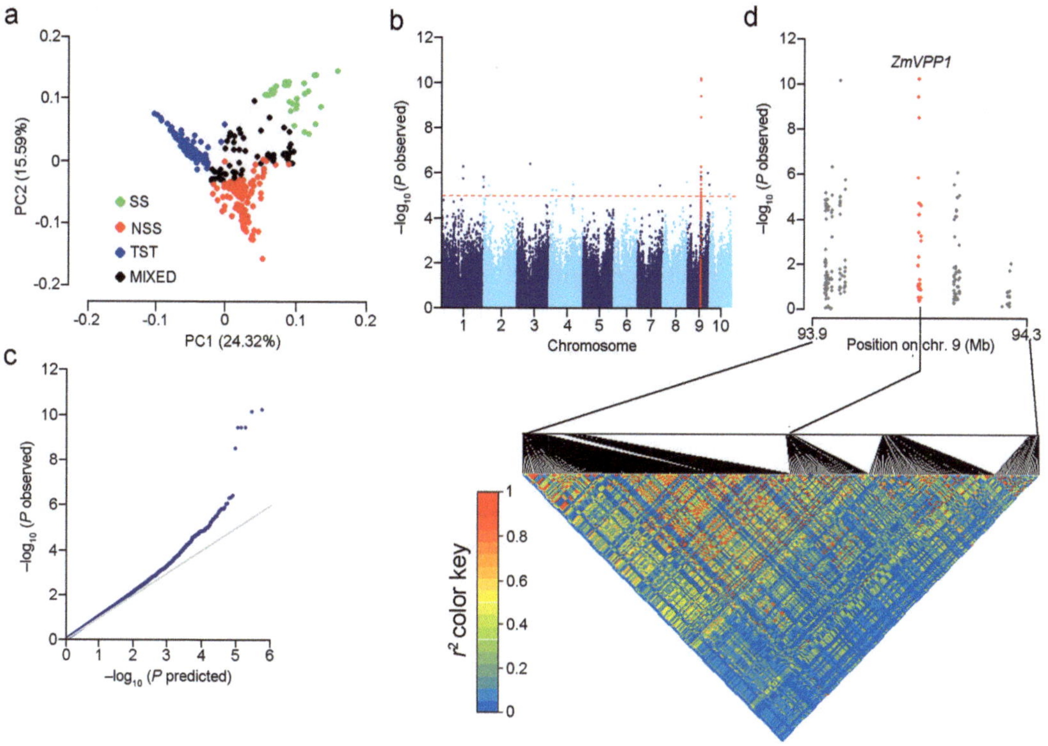

Fig. 1 A GWAS for drought resistance in maize at the seedling stage using a natural association panel comprising 368 maize inbred lines. (**a**) PCA plot of the first two components of the association panel. (**b**) Manhattan plot of the GWAS. The dashed horizontal line depicts the Bonferroni-adjusted significance threshold ($p = 1.0 \times 10^{-5}$). SNPs residing within the 200 kb genomic region on either side of the most significant SNP are shown in red. (**c**) Quantile–quantile plot for GWAS data using a mixed linear model. (**d**) Regional Manhattan plot of the *ZmVPP1* genomic region on chromosome 9 and an LD heatmap

3. Create the quantile-quantile (QQ) plot, in which $-\log_{10}$(observed p values) are plotted against the $-\log_{10}$(predicted p values) under the null hypothesis (Fig. 1c).
4. Identify the most significant SNPs located within *ZmVPP1* residing on chromosome 9 (Fig. 1d). More detailed interpretation of the results can be found in [10].

4 Notes

1. Trimmomatic (Version 0.39; http://www.usadellab.org/cms/?page=trimmomatic) performs a variety of useful trimming tasks on Illumina paired-end and single-end sequencing data. The selection of steps and parameters are provided in the command line [35]. FASTX-Toolkit (http://hannonlab.cshl.edu/fastx_toolkit/commandline.html) can also be used to preprocess the FASTA/FASTQ files before mapping the sequences to a reference genome.

2. BWA (http://bio-bwa.sourceforge.net/bwa.shtml) is a software program used to map low-divergent sequences to a large reference genome, utilizing three algorithms: BWA-backtrack, BWA-SW, and BWA-MEM. BWA-MEM is generally recommended for high-quality queries as it is faster and more accurate. HISAT2 is also an efficient and less memory-usage alignment program for mapping next-generation sequencing reads to a population of human genomes, as well as to a single reference genome.
3. EINGENSTRAT and STRUCTURE software programs are commonly used to infer population structure and to assign germplasms to subpopulations. The EINGENSTRAT method estimates the PCs from genotype data, and then includes the top 1–10 PCs as covariates in a generalized linear model [5]. The program STRUCTURE includes inferring distinct populations, assigning individuals to populations, studying hybrid zones, identifying migrants and admixed individuals, and estimating population allele frequencies in situations where many individuals are migrants or admixed [36].
4. SPAGeDi and Tassel software can be used to estimate and construct a kinship matrix among individuals [37].
5. Bioteke RNA extraction kit (Bioteke, Beijing, China), Superscript II Transcription Kit (Invitrogen), and reagents that Illumina supplied according to standard protocol were used for library construction and sequencing. The sequencing data for this chapter have been deposited in the NCBI Sequence Read Archive under accession code SRP026161 (https://www.ncbi.nlm.nih.gov/sra/?term=%20SRP026161) [33].
6. The phenotype in maize seedlings was provided for a panel of 368 maize inbred lines [10].
7. Seedling survival rate under severe water deficits was provided for a panel of 368 maize inbred lines [9, 10, 17, 18]. Components of grain yield that are often indexed as days to anthesis, days to silking, anthesis-silking interval, grain yield, ear-related phenotypes, and so on [12, 19–27]. Root traits that are often indexed as total number of roots, root dry weight, convex root area, primary root length, and so on [11, 28–32]. Variation in gene expression level can also be regarded as an expression trait (e-trait), and may be fundamental to stress response and adaptation [16].
8. Three populations have been identified and analyzed. Natural populations, with natural evolution and a complex breeding history, have undoubtedly created both population structure and complex familial relationships in both maize [38]. Multiparent populations, nested association mapping (NAM) and multiparent advanced generation intercross (MAGIC)

methodologies, have been used to overcome the problem of population structure and improve the statistical power for detecting rare variations [39, 40]. Random-open-parent association mapping (ROAM) populations allow direct integration of new populations into currently existing population resources for large-scale genetic analysis [41].

9. Seedling survival assay in large soil-containing blocks. The block is divided into enough plots to accommodate all the different germplasms in the assay with no fewer than two replicates to estimate the random errors.
10. Another approach is to conduct a drought test under field conditions using a naturally dry area or an area that is under management of controlled irrigation. Germplasms are grown in the field to maturity. Managed irrigation or naturally occurring drought is used to stress the plants and the phenotypic response of the plants is assessed. A number of agronomic traits can be evaluated in this approach, including grain yield.
11. Another approach is the utilization of osmotic stress experiments. Membrane-impermeable compounds (e.g., polyethylene glycol) are used to establish levels of osmotic stress that are equivalent to different levels of soil water deficit.
12. Early types of molecular markers developed, such as restriction fragment length polymorphisms (RFLPs), random amplified polymorphism DNA (RAPD), sequence characterized amplified regions, and simple sequence repeats (SSRs), rely on DNA hybridization or polymerase chain reaction (PCR) techniques [42].
13. High-throughput sequencing technology could obtain millions of SNP markers at a comparatively low cost for a mapping population. The primary molecular markers which are currently utilized have changed from DNA fragment lengths to the level of a single nucleotide polymorphism (SNP). RNA-seq, array-based genotyping methods, and genotyping by sequencing (GBS), are now often used in organisms with large genomes [14, 43].
14. Successful association mapping of populations depends on genetic diversity and the extent of LD decay in the mapping population, which determine the resolution and power of the mapping.
15. Q is an $n \times p$ matrix, where n is the number of germplasms and p is the number of defined subpopulations. Random molecular markers throughout the genome are generally used to estimate Q with STRUCTURE software [36, 44]. Another successful approach in modeling complex population structure is to estimate admixture proportions based on clustering-based

methods available in ADMIXTURE software [45]. EINGEN STRAT software, based on principal component analysis can be used to estimate ancestry differences in a mapping population [46].

16. K is an $n \times n$ matrix, which represents a marker-based estimate of probability of identity by descent. SPAGeDi software is used to estimate the familial relatedness among the germplasms in a kinship matrix (K) [37]. TASSEL software is efficient in constructing a kinship matrix [47].
17. The basic statistics for association analysis are linear regression, analysis of variance (ANOVA), and a t-test or chi-square test [5]. Individuals always share complex ancestral relatedness in a natural population due to intercrossing efforts in crop breeding. Since population structure can generate spurious genotype–phenotype associations, different statistical approaches have been designed to deal with this confounding factor.
18. The GLM model used the following equation: $y = X\alpha + e$. In the equation, y represents phenotype, X represents genotype, α represents the genotype effect, and e represents random effects [48].
19. The Q and PCA models used the following equation: $y = X\alpha + P\beta + e$. In the equation, y represents phenotype, X represents genotype, α represents the genotype effect, P represents the first 2–5 PCs or Q matrix, β represents the PCs or Q effects, and e represents random effects [49].
20. The K model used the following equation: $y = X\alpha + K\mu + e$. In the equation, y represents phenotype, X represents genotype, α represents the genotype effect, K represents the relative kinship matrix, and μ and e represent random effects.
21. The $Q + K$ and PCA + K models used the following equation: $y = X\alpha + P\beta + K\mu + e$. In the equation, y represents phenotype, X represents genotype, α represents the genotype effect, P represents the first 2–5 PCs or Q matrix, β represents the PCs or Q effects, K represents the relative kinship matrix, and μ and e represent random effects [7, 9, 10, 13, 17, 33, 50, 51].
22. Data format conversion: vcftools --vcf test.vcf --plink --out name.plink; ./ run_pipeline.pl -fork1 -h test. hmp -export -exportType Plink..
23. The MLM model is not efficient for large data sets when associations are analyzed for a large number of markers with the phenotypic information from thousands of individuals. The population parameters previously determined (P3D) method has enabled the computational burden required for GWAS to be reduced sufficiently to run on a personal desktop computer.

Acknowledgments

The authors would like to thank Dr Hongwei Wang for the helpful discussion. This research was supported by the National Natural Science Foundation of China (31625022 and 31971952), Beijing Outstanding Young Scientist Program (BJJWZYJH01201910019026), and the National Key Research and Development Plan of China (2016YFD0100605).

References

1. Gupta A, Rico-Medina A, Cano-Delgado AI (2020) The physiology of plant responses to drought. Science 368(6488):266–269. https://doi.org/10.1126/science.aaz7614
2. Levitt J (1980) Responses of plants to environmental stresses. Volume II. Water, radiation, salt, and other stresses. Academic Press, London. 607 p
3. Yoshida T, Mogami J, Yamaguchi-Shinozaki K (2014) ABA-dependent and ABA-independent signaling in response to osmotic stress in plants. Curr Opin Plant Biol 21:133–139. https://doi.org/10.1016/j.pbi.2014.07.009
4. Liu S, Qin F (2021) Genetic dissection of maize drought tolerance for trait improvement. Mol Breed 41(2):8. https://doi.org/10.1007/s11032-020-01194-w
5. Yu J, Buckler ES (2006) Genetic association mapping and genome organization of maize. Curr Opin Biotechnol 17(2):155–160. https://doi.org/10.1016/j.copbio.2006.02.003
6. Huang X, Han B (2014) Natural variations and genome-wide association studies in crop plants. Annu Rev Plant Biol 65:531–551. https://doi.org/10.1146/annurev-arplant-050213-035715
7. Huang X, Zhao Y, Wei X et al (2011) Genome-wide association study of flowering time and grain yield traits in a worldwide collection of rice germplasm. Nat Genet 44(1):32–39. https://doi.org/10.1038/ng.1018
8. Jia G, Huang X, Zhi H et al (2013) A haplotype map of genomic variations and genome-wide association studies of agronomic traits in foxtail millet (*Setaria italica*). Nat Genet 45(8):957–961. https://doi.org/10.1038/ng.2673
9. Liu S, Wang X, Wang H et al (2013) Genome-wide analysis of *ZmDREB* genes and their association with natural variation in drought tolerance at seedling stage of *Zea mays* L. PLoS Genet 9(9):e1003790. https://doi.org/10.1371/journal.pgen.1003790
10. Wang X, Wang H, Liu S et al (2016) Genetic variation in *ZmVPP1* contributes to drought tolerance in maize seedlings. Nat Genet 48(10):1233–1241. https://doi.org/10.1038/ng.3636
11. Li X, Guo Z, Lv Y et al (2017) Genetic control of the root system in rice under normal and drought stress conditions by genome-wide association study. PLoS Genet 13(7):e1006889. https://doi.org/10.1371/journal.pgen.1006889
12. Bhandari A, Sandhu N, Bartholome J et al (2020) Genome-Wide Association Study for yield and yield related traits under reproductive stage drought in a diverse indica-aus rice panel. Rice 13(1):53. https://doi.org/10.1186/s12284-020-00406-3
13. Huang X, Wei X, Sang T et al (2010) Genome-wide association studies of 14 agronomic traits in rice landraces. Nat Genet 42(11):961–967. https://doi.org/10.1038/ng.695
14. Chia JM, Song C, Bradbury PJ et al (2012) Maize HapMap2 identifies extant variation from a genome in flux. Nat Genet 44(7):803–807. https://doi.org/10.1038/ng.2313
15. Xu J, Chen G, Hermanson PJ et al (2019) Population-level analysis reveals the widespread occurrence and phenotypic consequence of DNA methylation variation not tagged by genetic variation in maize. Genome Biol 20(1):243. https://doi.org/10.1186/s13059-019-1859-0
16. Liu S, Li C, Wang H et al (2020) Mapping regulatory variants controlling gene expression in drought response and tolerance in maize. Genome Biol 21(1):163. https://doi.org/10.1186/s13059-020-02069-1
17. Mao H, Wang H, Liu S et al (2015) A transposable element in a *NAC* gene is associated with drought tolerance in maize seedlings. Nat Commun 6:8326. https://doi.org/10.1038/ncomms9326
18. Xiang Y, Sun X, Gao S et al (2017) Deletion of an endoplasmic reticulum stress response

element in a *ZmPP2C-A* gene facilitates drought tolerance of maize seedlings. Mol Plant 10(3):456–469. https://doi.org/10.1016/j.molp.2016.10.003

19. Lu Y, Zhang S, Shah T et al (2010) Joint linkage-linkage disequilibrium mapping is a powerful approach to detecting quantitative trait loci underlying drought tolerance in maize. Proc Natl Acad Sci U S A 107(45): 19585–19590. https://doi.org/10.1073/pnas.1006105107
20. Xue Y, Warburton ML, Sawkins M et al (2013) Genome-wide association analysis for nine agronomic traits in maize under well-watered and water-stressed conditions. Theor Appl Genet 126(10):2587–2596. https://doi.org/10.1007/s00122-013-2158-x
21. Thirunavukkarasu N, Hossain F, Arora K et al (2014) Functional mechanisms of drought tolerance in subtropical maize (*Zea mays* L.) identified using genome-wide association mapping. BMC Genomics 15(1):1182–1182. https://doi.org/10.1186/1471-2164-15-1182
22. Farfan ID, De La Fuente GN, Murray SC et al (2015) Genome wide association study for drought, aflatoxin resistance, and important agronomic traits of maize hybrids in the sub-tropics. PLoS One 10(2):e0117737. https://doi.org/10.1371/journal.pone.0117737
23. Pantaliao GF, Narciso M, Guimaraes C et al (2016) Genome wide association study (GWAS) for grain yield in rice cultivated under water deficit. Genetica 144(6): 651–664. https://doi.org/10.1007/s10709-016-9932-z
24. Ma X, Feng F, Wei H et al (2016) Genome-Wide Association Study for plant height and grain yield in rice under contrasting moisture regimes. Front Plant Sci 7:1801. https://doi.org/10.3389/fpls.2016.01801
25. Li L, Mao X, Wang J et al (2019) Genetic dissection of drought and heat-responsive agronomic traits in wheat. Plant Cell Environ 42(9):2540–2553. https://doi.org/10.1111/pce.13577
26. Gahlaut V, Jaiswal V, Singh S et al (2019) Multi-locus genome wide association mapping for yield and its contributing traits in hexaploid wheat under different water regimes. Sci Rep 9(1):19486. https://doi.org/10.1038/s41598-019-55520-0
27. Abou-Elwafa SF, Shehzad T (2021) Genetic diversity, GWAS and prediction for drought and terminal heat stress tolerance in bread wheat (*Triticum aestivum* L.). Genet Resour Crop Evol 68(2):711–728. https://doi.org/10.1007/s10722-020-01018-y
28. Pace J, Gardner C, Romay C et al (2015) Genome-wide association analysis of seedling root development in maize (*Zea mays* L.). BMC Genomics 16(1):47. https://doi.org/10.1186/s12864-015-1226-9
29. Phung NT, Mai CD, Hoang GT et al (2016) Genome-wide association mapping for root traits in a panel of rice accessions from Vietnam. BMC Plant Biol 16:64. https://doi.org/10.1186/s12870-016-0747-y
30. Zaidi PH, Seetharam K, Krishna G et al (2016) Genomic regions associated with root traits under drought stress in tropical maize (*Zea mays* L.). PLoS One 11(10):e0164340. https://doi.org/10.1371/journal.pone.0164340
31. Zhang X, Mi Y, Mao H et al (2020) Genetic variation in *ZmTIP1* contributes to root hair elongation and drought tolerance in maize. Plant Biotechnol J 18(5):1271–1283. https://doi.org/10.1111/pbi.13290
32. Guo J, Li C, Zhang X et al (2020) Transcriptome and GWAS analyses reveal candidate gene for seminal root length of maize seedlings under drought stress. Plant Sci 292:110380. https://doi.org/10.1016/j.plantsci.2019.110380
33. Fu J, Cheng Y, Linghu J et al (2013) RNA sequencing reveals the complex regulatory network in the maize kernel. Nat Commun 4: 2832. https://doi.org/10.1038/ncomms3832
34. Yu J, Pressoir G, Briggs WH et al (2006) A unified mixed-model method for association mapping that accounts for multiple levels of relatedness. Nat Genet 38(2):203–208. https://doi.org/10.1038/ng1702
35. Bolger AM, Lohse M, Usadel B (2014) Trimmomatic: a flexible trimmer for Illumina sequence data. Bioinformatics 30(15): 2114–2120. https://doi.org/10.1093/bioinformatics/btu170
36. Pritchard JK, Stephens M, Donnelly P (2000) Inference of population structure using multilocus genotype data. Genetics 155(2):945–959
37. Hardy OJ, Vekemans X (2002) spagedi: a versatile computer program to analyse spatial genetic structure at the individual or population levels. Mol Ecol Notes 2(4):618–620. https://doi.org/10.1046/j.1471-8286.2002.00305.x
38. Flint-Garcia SA, Thuillet AC, Yu J et al (2005) Maize association population: a high-resolution platform for quantitative trait locus dissection. Plant J 44(6):1054–1064. https://doi.org/10.1111/j.1365-313X.2005.02591.x

39. Cavanagh C, Morell M, Mackay I et al (2008) From mutations to MAGIC: resources for gene discovery, validation and delivery in crop plants. Curr Opin Plant Biol 11(2):215–221. https://doi.org/10.1016/j.pbi.2008.01.002
40. Buckler ES, Holland JB, Bradbury PJ et al (2009) The genetic architecture of maize flowering time. Science 325(5941):714–718. https://doi.org/10.1126/science.1174276
41. Xiao Y, Tong H, Yang X et al (2016) Genome-wide dissection of the maize ear genetic architecture using multiple populations. New Phytol 210(3):1095–1106. https://doi.org/10.1111/nph.13814
42. Kim NS, Park NI, Kim SH et al (2000) Isolation of TC/AG repeat microsatellite sequences for fingerprinting rice blast fungus and their possible horizontal transfer to plant species. Mol Cell 10(2):127–134. https://doi.org/10.1007/s10059-000-0127-0
43. Wang W, Mauleon R, Hu Z et al (2018) Genomic variation in 3,010 diverse accessions of Asian cultivated rice. Nature 557(7703):43–49. https://doi.org/10.1038/s41586-018-0063-9
44. Falush D, Stephens M, Pritchard JK (2003) Inference of population structure using multilocus genotype data: linked loci and correlated allele frequencies. Genetics 164(4):1567–1587
45. Alexander DH, Novembre J, Lange K (2009) Fast model-based estimation of ancestry in unrelated individuals. Genome Res 19(9):1655–1664. https://doi.org/10.1101/gr.094052.109
46. Price AL, Patterson NJ, Plenge RM et al (2006) Principal components analysis corrects for stratification in genome-wide association studies. Nat Genet 38(8):904–909. https://doi.org/10.1038/ng1847
47. Bradbury PJ, Zhang Z, Kroon DE et al (2007) TASSEL: software for association mapping of complex traits in diverse samples. Bioinformatics 23(19):2633–2635. https://doi.org/10.1093/bioinformatics/btm308
48. Wang Y, Tao Z, Wang W et al (2020) Molecular variation in a functionally divergent homolog of FCA regulates flowering time in *Arabidopsis thaliana*. Nat Commun 11(1):5830. https://doi.org/10.1038/s41467-020-19666-0
49. Thornsberry JM, Goodman MM, Doebley J et al (2001) Dwarf8 polymorphisms associate with variation in flowering time. Nat Genet 28(3):286–289. https://doi.org/10.1038/90135
50. Li H, Peng Z, Yang X et al (2013) Genome-wide association study dissects the genetic architecture of oil biosynthesis in maize kernels. Nat Genet 45(1):43–50. https://doi.org/10.1038/ng.2484
51. Liu H, Luo X, Niu L et al (2017) Distant eQTLs and Non-coding sequences play critical roles in regulating gene expression and quantitative trait variation in maize. Mol Plant 10(3):414–426. https://doi.org/10.1016/j.molp.2016.06.016

Chapter 17

Chemical Approaches for Improving Plant Water Use

Aditya S. Vaidya and Sean R. Cutler

Abstract

Agricultural productivity in rain-fed crops has been threatened in recent decades due to increased instances of drought and diminishing freshwater resources. This has led to the development of novel chemical and genetic approaches for improving plant water use efficiency. Agrochemical water-banking with the aid of synthetic mimics of phytohormone abscisic acid (ABA) is one such approach, whereby plant transpiration can be chemically tuned to ensure water availability during critical stages of growth. Here, we describe the use of infrared thermography, a noninvasive quantitative technique to evaluate antitranspirant efficacy of existing ABA receptor agonists in crops such as wheat and tomato.

Key words Abscisic acid, Opabactin, Agonist, Drought, Infrared thermography, Antitranspirants

1 Introduction

With an ever-increasing global population, estimated to reach 11 billion by the end of the century, it has become imperative to improve upon global crop productivity. Mitigating the effects of drought on plant water use is critical to boosting crop productivity, which if left unchecked can result in billions of dollars in agricultural losses annually [1]. Abscisic acid (ABA) (Fig. 1a) is a key stress hormone found in plants that regulates guard cell closure, in turn reducing plant transpiration and aiding plant acclimatization to drought stress [2]. However, guard cell closure also limits CO_2 intake, reducing photosynthetic efficiency, leading to an unwanted trade-off between water use and growth. This trade-off makes it intrinsically difficult to incorporate both traits of high water use efficiency and high yield in economically relevant crops, though some strides have been made in this direction by Monsanto's DroughtGard™ that is reported to boost yields in maize by ≈6% under moderate drought by overexpression of a Bacillus subtilis cold-stress induced RNA chaperon protein [3]. Application of synthetic mimics of ABA is an attractive alternative to traditional breeding techniques as it can coerce diverse high yielding crops into

Takuya Yoshida (ed.), *Abscisic Acid: Methods and Protocols*,
Methods in Molecular Biology, vol. 2462, https://doi.org/10.1007/978-1-0716-2156-1_17,

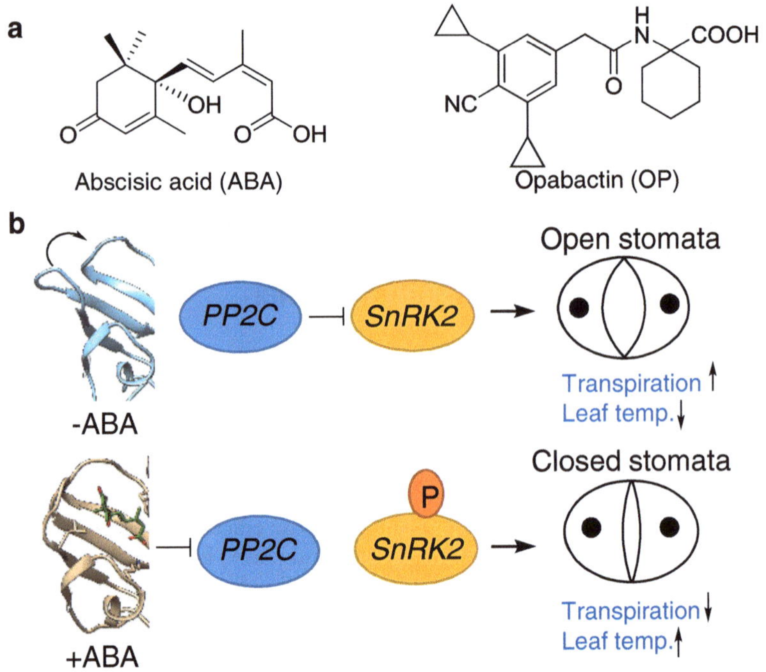

Fig. 1 (**a**) Structures of *S*-(+) abscisic acid (ABA) and opabactin (OP). (**b**) Biochemical steps following activation of ABA receptors and guard cell response in presence of ABA or ABA mimics and its effect on transpiration and overall leaf temperature

more water efficient states "on-demand," reserving water for critical flowering and grain production stages [4]. This agrochemical "water-banking" strategy combined with data inputs from "precision agriculture" can potentially save and optimally utilize limited water resources in face of adverse climatic conditions.

The identification of ABA receptors in 2009 [5, 6] and recent advances in understanding of the core ABA signaling pathway [7] has propelled the development of numerous synthetic mimics of ABA as ABA itself is metabolically labile, undergoes photo degradation and is expensive, limiting its broad scale application on the field [8]. To date, numerous ABA receptor agonists [8–23] have been described by industry and academia that function via mimicking the conformational changes brought about by ABA on the soluble PYRABACTIN RESISTANCE 1/PYR1-LIKE/REGU LATORY COMPONENT OF ABA RECEPTOR (hereafter PYR/PYL) ABA receptors. ABA bound PYR/PYL receptor complex inhibits the function of clade A type 2C protein phosphatases (PP2Cs) in turn, leading to accumulation of their phosphorylated substrates-SNF1-related protein kinase 2s (SnRK2s) (Fig. 1b), which phosphorylate other downstream factors resulting in guard cell closure amongst other physiological responses [24, 25]. This module is thus an attractive target for agrochemical control of plant water use. Recently, an overpowered ABA agonist-Opabactin (OP) [26] (Fig. 1a) was reported by us that had superior potency and antitranspirant activity as compared to other reported agonists such as quinabactin [27] and AMF4 [28] in economically important crops such as wheat and tomato.

Here, we describe the application and evaluation of antitranspirant efficacy of OP and ABA using quantitative infra-red thermography [29] in crops such as wheat and tomato. This technique precisely reports elevation in leaf surface temperature as a consequence of reduced evaporative cooling when transpiration is arrested by ABA mimics (Fig. 1b).

2 Materials

All solutions are prepared in ultrapure water (sensitivity of 18 MΩ cm at 25 °C). All other chemicals are reagent grade, unless specified otherwise.

2.1 Chemicals

1. OP: The chemical synthesis of OP can be performed (*see* **Note 1**). OP is also commercially available from vendors (e.g., Kerafast). Stored at −20 °C and stocks prepared in ACS reagent grade dimethyl sulfoxide (DMSO) at 10–50 mM concentration, typically aliquoted in smaller volumes (100–200 μL) in clear Eppendorf tubes and stored at −20 °C until further use.
2. ABA: (*S*)-(+) Abscisic acid. Stored at −20 °C and stocks prepared in ACS reagent grade DMSO at 10–50 mM concentration, typically aliquoted in smaller volumes (100–200 μL) in clear Eppendorf tubes and stored at −20 °C until further use. ABA is photosensitive and should be stored away from light.
3. Silwet L-77.
4. Deionized water.

2.2 Plant Materials and Growth Media

1. Tomato seeds (UC82).
2. Wheat seeds (Cal Rojo).
3. 15 mm sterile polystyrene petri dishes.
4. 1.5 mL clear Eppendorf tubes.
5. 50 mL falcon tubes.
6. Tube rotator/revolver.
7. 1 L beakers.
8. Seed germination media: ½ MS sucrose, 0.7% agar media. To 800 mL of deionized water in a 1 L glass bottle is added 1.73 g of Murashige and Skoog Basal Salts, 4 g of sucrose and 5.6 g of agar and the mixture stirred to dissolve the sucrose and MS salts and suspend the agar (*see* **Note 2**). Sterilized media is stored at

room temperature until further use (*see* **Note 3**). Approximately 20 mL of media (55 °C) is poured onto 15 mm sterile polystyrene petri dishes and allowed to solidify under the flow hood.

9. 70% ethanol/water.
10. 10% Bleach with 0.01% Tween 20 in deionized water.
11. Growth chamber.
12. 4 in. plastic pots filled with soil.

2.3 Infra-Red Thermography

1. Infra-red Camera (e.g., FLIR T62101 or equivalent).
2. Trays or receptacle.
3. Spray bottles.
4. Image analysis software (e.g., FLIR Tools software version 5.215161.1001 or equivalent).
5. Statistical analysis software (e.g., Graph Pad Prism 6.0 and Microsoft Excel).

3 Methods

3.1 Preparation of Plant Material

1. 30–50 tomato seeds (UC-82) or 100–150 wheat seeds (WB9229) are taken in a 50 mL falcon tube and rinsed with 30 mL of 70% ethanol/water for 2 min by gentle rotation on a tube rotator/revolver at room temperature.
2. The seeds are allowed to settle and the supernatant ethanol discarded. Thereafter, the seeds are exposed to 30 mL of 10% Bleach with 0.01% Tween 20 in deionized water and gently rotated for 45 min at room temperature.
3. The seeds are allowed to settle and the supernatant bleach solution discarded. Diligently follow all waste disposal regulations when disposing waste materials.
4. Seeds are then washed 3 times with 30 mL of deionized water. The seeds (20–40) are now transferred and spread on media containing ½ MS sucrose, 0.7% agar in 15 mm polystyrene petri dishes.
5. The petri dishes are stored in the dark at room temperature until radicle emergence is observed (7–10 days—*see* **Note 4**).
6. Tomato seedlings are transferred in a growth chamber at 22 °C under 16 h day/8 h night illumination for 4–5 days to initiate cotyledon greening and expansion. Tomato seedlings (3–5 cm in radicle length) with expanded green cotyledons are transferred to soil in 4-in. plastic pots (1 seedling/pot—placed centrally). The seedlings are grown in a growth room at 22 °C under 16-h day/8 h night illumination for 4–5 days

(*see* **Note 5**) before being transferred under LED illumination at 120–150 μmol/m^2/s photosynthetic active radiation for 3–4 weeks at 25 °C.

7. In the case of wheat, germinated seedlings are directly transferred to soil (10–15 seedlings) in 4-in. plastic pots. Seedlings should be uniformly distributed in the soil and grown under 120–150 μmol/m^2/s photosynthetic active radiation (*see* **Notes 6** and 7) for 2–3 weeks at 22 °C.

3.2 Preparation of Chemical Solutions

1. Weigh the appropriate amounts of chemicals (*see* **Note 8**) in 1.5 mL clear Eppendorf tubes to make at least 1000 μL stock at 10 mM in DMSO.
2. Take 1 L of ultrapure water in a beaker and add 500 μL of Silwet L-77 as a wetting agent to achieve a final concentration of 0.05% v/v. Stir the solution for 15 min at room temperature, to completely dissolve the surfactant in water. If the solution appears turbid, discard and use freshly received Silwet.
3. A 10 mL solution of 50 μM of chemicals in Silwet containing water is prepared just before aerosolized application on plants, by adding 10 mL of Silwet solution to 50 μL of 10 mM stock in a clean 50 mL falcon tube (final concentration of DMSO 0.5% v/v).

3.3 Preparation of Plants Before Aerosolized Chemical Application

1. Pot positions within each experiment are randomized (using an Excel function) to control for micro environmental differences in the growth chamber.
2. The plants are uniformly placed so that the aerial parts of one plant do not touch its immediate neighbors.

3.4 Application of Aerosolized Chemicals on Plants

1. Solutions of ABA agonists are prepared just before application as described in the preceding sections.
2. Pots are placed in a tray/receptacle and 20 mL of 50 μM chemical solution in case of tomato and 10 mL of 50 μM chemical solution in case of wheat or mock control containing 0.5% v/v DMSO is applied in the form of aerosols with the help of spray bottles (*see* **Notes 9** and **10**), ensuring uniform exposure to all aerial parts of the plants.
3. The pots are then placed back in randomized positions, uniformly spaced in flats of sufficient size so that aerial parts of plants treated with different chemicals/mock do not touch each other.
4. Water is added to reach soil field capacity throughout the duration of the experiment.

3.5 Recording of Infrared Images

1. Infrared images of chemically treated plants are recorded using the infra-red camera (e.g., FLIR T62101 Infrared camera or comparable models). The camera should be set up (*see* **Note**

11) and turned on a tripod placed in the growth chamber used for the experiment at least 15–20 min before recording infrared images.

2. Infrared images of tomato plants are taken in settings where the plant is placed below the camera at a suitable distance to ensure proper focus (typically autofocus option is used) and maximum coverage of the camera frame. In the case of wheat, the camera should be placed at a suitable distance to record the leaf sheaths across the frame. The best settings can be achieved with dummy plants not involved in the experiment.
3. Once the best settings have been achieved, infrared images of plants are recorded in minimum time, avoiding bias toward any particular treatment group (*see* **Note 12**).

3.6 Processing of Infrared Images

1. IR images are processed using appropriate software (e.g., FLIR Tools software).
2. Using the software, the temperature scales are modified on all images uniformly without any bias toward any particular treatment group. Typically, 0.5–2 °C change in leaf temperature is obtained relative to untreated controls depending on the type of chemical, concentration at which it is treated, the duration of treatment and the species used for the experiment.
3. Average leaf temperatures are recorded from different treatments by averaging 10–15 ~1 cm^2 leaf area spots per replicate pot, recorded in Microsoft Excel sheets. Leaf area spots should be uniform in size to prevent any bias toward any particular area of the leaf that may be hot or cold simultaneously covering the entire leaf area.

3.7 Data Analysis and Presentation

1. Statistical analysis is conducted using appropriate software (e.g., GraphPad Prism 6.0). The mean leaf temperatures are used in post-hoc Dunnett test comparing mock to treatment effects, providing multiplicity adjusted *p*-values for treatment effects.
2. Final presentation of the data usually involves presenting infrared images recorded for each treatment including corresponding bar graphs (Fig. 2a–d). Temperature scales must be shown that indicate the leaf temperature.

4 Notes

1. The chemical synthesis of OP can be performed as per synthetic schemes [26].
2. Typically, ½ MS salts and sucrose are added to the water first and stirred to ensure complete dissolution. If a clear solution is not obtained, discard and use freshly received materials. Since

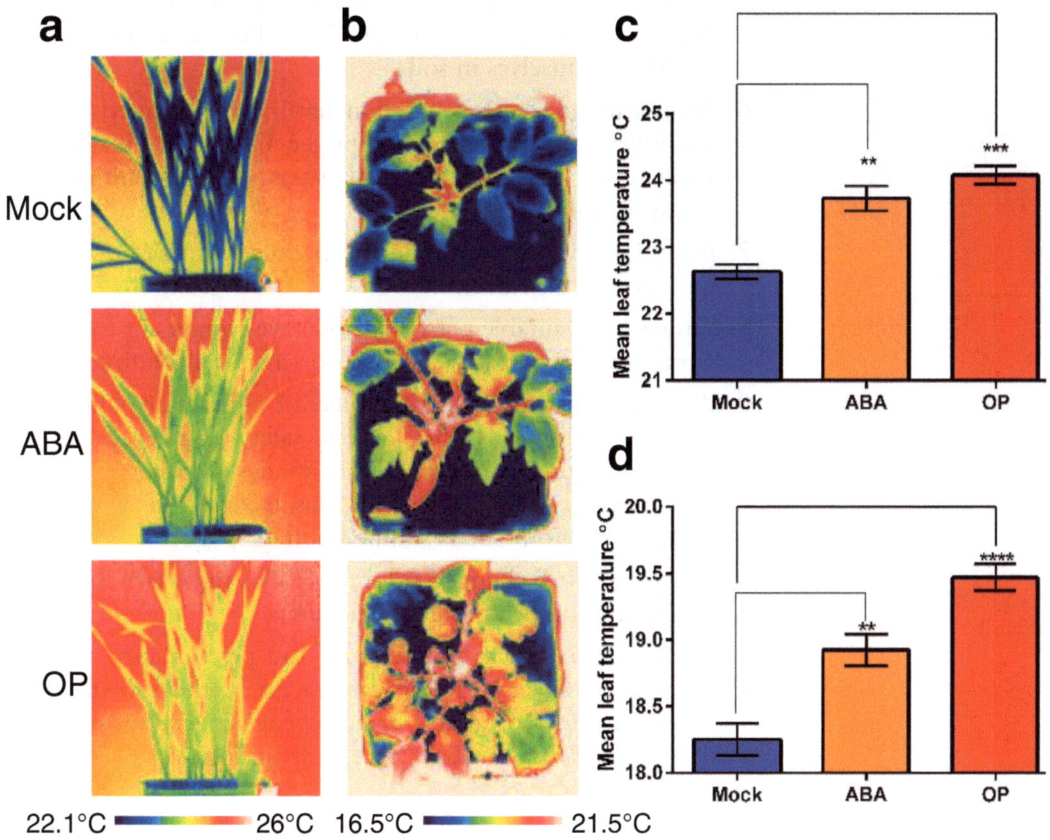

Fig. 2 (**a**) Infrared images of 2–3-week-old wheat seedlings (Cal Rojo) treated with 50 μM solutions composed of 0.5% v/v DMSO and 0.05% Silwet-77. Graphs depict the leaf temperatures of different treatments relative to mock-treated plants at 24 h ($n = 3$ replicated for mock and $n = 3$ for chemical treatment). (**b**) Infrared images of 3–4 week-old tomato seedlings (UC82) treated with 50 μM solutions composed of 0.5% v/v DMSO and 0.05% Silwet 77. Graphs depict the leaf temperatures of different treatments relative to mock-treated plants at 24 h ($n = 8$ replicated for mock and $n = 4$ for chemical treatment). (**c**) Quantification of leaf temperature in wheat seedlings as described in (**a**). (**d**) Quantification of leaf temperature in tomato seedlings as described in (**b**). Statistical analyses were conducted in GraphPad Prism 6.0 using one-way analysis of variance (ANOVA) and post hoc Dunnett test to obtain multiplicity adjusted p values of treatment effects relative to mock-treated controls

agar flakes are insoluble in water at room temperature and remain in suspension, adding them last to the media is ideal as it ensures other materials have completely dissolved. Media so obtained is not pH stabilized and used as is after sterilization by autoclaving.

3. Fresh media is always preferred for all experiments.
4. Prolonged storage of seeds may diminish their viability and subsequent germination rate, hence freshly harvested seeds are always recommended for experiments. Alternatively, seed germination rate of commercial varieties may be checked in preliminary germination experiments.

5. Seedlings freshly transferred from media to soil may be stressed and should be grown in high humidity for a few days to establish themselves in soil.
6. Photosynthetic active radiation levels are measured using a quantum PAR meter (e.g., Apogee Quantum Flux meter); avoid using plants that show signs of stress-anthocyanin accumulation.
7. Plants included in the intended experiment should be similar in size and leaf distribution.
8. Always wear suitable personal protective equipment to prevent direct contact with any chemicals or chemical solutions.
9. Wear appropriate personal protective equipment when conducting applications (i.e., gloves, safety goggles, lab coat). Avoid inhalation of aerosolized liquids by either wearing a respirator or conducting applications in conditions with appropriate ventilation (such as fume hood or spray chamber).
10. Discard residual chemical solutions, following established environmental and health safety guidelines.
11. The thermal camera temperature and performance can take time to stabilize; turn the camera on in the growth room ~15 min prior to taking images. Avoid collecting images in areas with heavy airflow (i.e., near airflow vents) to avoid artifacts.
12. Water from pots or flats that falls on leaf surfaces can affect their temperature; handle pots and flats carefully prior to collecting leaf temperature data.

Acknowledgments

Tomato (UC-82) and wheat (Cal Rojo) varieties used for thermography experiments were gifts from Prof. Linda Walling, University of California, Riverside and Prof. Adam Lukaszewsky, University of California, Riverside, respectively.

References

1. Nations FAAO of TU and Food and Agriculture Organization of the United Nations (2018) Tracking adaptation in agricultural sectors. FAO, Rome. https://doi.org/10.18356/87fe25de-en
2. Daszkowska-Golec A (2016) The role of abscisic acid in drought stress: how ABA helps plants to cope with drought stress. In: Hossain MA, Wani SH, Bhattacharjee S et al (eds) Drought stress tolerance in plants, Vol 2: Molecular and genetic perspectives. Springer International Publishing, Cham, pp 123–151
3. Nemali KS, Bonin C, Dohleman FG et al (2015) Physiological responses related to increased grain yield under drought in the first biotechnology-derived drought-tolerant maize. Plant Cell Environ 38:1866–1880
4. Çakir R (2004) Effect of water stress at different development stages on vegetative and

reproductive growth of corn. Field Crop Res 89:1–16

5. Park S-Y, Fung P, Nishimura N et al (2009) Abscisic acid inhibits type 2C protein phosphatases via the PYR/PYL family of START proteins. Science 324:1068–1071
6. Ma Y, Szostkiewicz I, Korte A et al (2009) Regulators of PP2C phosphatase activity function as abscisic acid sensors. Science 324: 1064–1068
7. Cutler SR, Rodriguez PL, Finkelstein RR et al (2010) Abscisic acid: emergence of a core signaling network. Annu Rev Plant Biol 61: 651–679
8. Helander JDM, Vaidya AS, Cutler SR (2016) Chemical manipulation of plant water use. Bioorg Med Chem 24:493–500
9. Frackenpohl J, Hans-Joachim Zeiss I, Heinemann I et al (2015) Use of substituted isoquinolinones, isoquinolindiones, isoquinolintriones and dihydroisoquinolinones or in each case salts thereof as active agents against abiotic stress in plants. https://patentimages.storage.googleapis.com/85/bf/14/6a7bab7007eff7/US9173395.pdf
10. Cutler SR, Wendeborn SV, Jung PJ et al (2016) Compounds that induce aba responses. https://patentimages.storage.googleapis.com/3c/8e/a5/53ecd16d2042a7/US20160280651A1.pdf
11. Cutler SR and Okamoto M (2016) Synthetic compounds for vegetative ABA responses. https://patentimages.storage.googleapis.com/0d/77/e5/b485cd4b65b762/US9345245.pdf
12. Dejonghe W, Okamoto M, Cutler SR (2018) Small molecule probes of ABA biosynthesis and signaling. Plant Cell Physiol 59:1490–1499
13. Lachia MD, Wendeborn SV, Jung PJM et al (2018) 2-oxo-3,4-dihydroquinoline compounds as plant growth regulators. https://patentimages.storage.googleapis.com/02/4b/ee/fc4a6a0f677484/US20180044297A1.pdf
14. Frackenpohl J, Bojack G, Helmke H et al (2017) Use of substitute oxo tetrahydroquinoline sulfonamides or salts thereof for raising stress tolerance of plants. https://patentimages.storage.googleapis.com/0d/e0/59/3fb3cd32900682/US20170027172A1.pdf
15. Frackenpohl J, Bojack G, Helmke H et al (2016) Use of substituted dihydrooxindolylsulfonamides, or the salts thereof, for increasing the stress tolerance of plants. https://patentimages.storage.googleapis.com/cb/6e/25/842328f2b2cbef/US20160237035A1.pdf
16. Frackenpohl J, Bojack G, Helmke H et al (2018) Substituted 1-cycloalkyl-2-oxotetrahydroquinolin-6-ylsulfonamides or salts thereof and use thereof to increase stress tolerance in plants. https://patentimages.storage.googleapis.com/6d/43/e5/67dd0ad006b0f2/US20180020662A1.pdf
17. Frackenpohl J, Willms L, and Dittgen J (2017) Substituted cyano cycloalkyl penta-2, 4-dienes, cyano cycloalkyl pent-2-en-4-ynes, cyano heterocyclyl penta-2, 4-dienes and cyano heterocyclyl pent-2-en-4-ynes as https://patents.google.com/patent/US20170210701A1/en
18. Frackenpohl J, Heinemann I, Müller T et al (2011) Aryl-and hetarylsulfonamides as active ingredients against abiotic plant stress. https://patentimages.storage.googleapis.com/a7/2d/ea/8a60a636e4fdd7/US20110230350A1.pdf
19. Cutler SR, Lachia MD, Wendeborn SV et al (2018) Carbamate quinabactin. https://patentimages.storage.googleapis.com/65/27/d8/68509d60cc0199/WO2018017490A1.pdf
20. Godfrey CRA, Lachia MD, Wendeborn SV et al (2018) Plant growth regulator compounds. https://patentimages.storage.googleapis.com/87/fd/61/816a6c284d84f1/WO2018007217A1.pdf
21. Cutler SR, Wendeborn SV, Loiseleur O et al (2018) Derivatives of halo quinabactin. https://patentimages.storage.googleapis.com/af/59/c9/c06faa2b1d5db6/US20180312470A1.pdf
22. Vaidya AS, Peterson FC, Yarmolinsky D et al (2017) A rationally designed agonist defines subfamily IIIA abscisic acid receptors as critical targets for manipulating transpiration. ACS Chem Biol 12:2842–2848
23. Elzinga D, Sternburg E, Sabbadin D et al (2019) Defining and exploiting hypersensitivity hotspots to facilitate abscisic acid agonist optimization. ACS Chem Biol 14:332–336
24. Weiner JJ, Peterson FC, Volkman BF et al (2010) Structural and functional insights into core ABA signaling. Curr Opin Plant Biol 13: 495–502
25. Miyakawa T, Fujita Y, Yamaguchi-Shinozaki K et al (2013) Structure and function of abscisic acid receptors. Trends Plant Sci 18:259–266
26. Vaidya AS, Helander JDM, Peterson FC et al (2019) Dynamic control of plant water use using designed ABA receptor agonists. Science 366:446–454

27. Okamoto M, Peterson FC, Defries A et al (2013) Activation of dimeric ABA receptors elicits guard cell closure, ABA-regulated gene expression, and drought tolerance. Proc Natl Acad Sci U S A 110:12132–12137
28. Cao M-J, Zhang Y-L, Liu X et al (2017) Combining chemical and genetic approaches to increase drought resistance in plants. Nat Commun 8:1183
29. Costa JM, Grant OM, Chaves MM (2013) Thermography to explore plant–environment interactions. J Exp Bot 64:3937–3949

INDEX

Takuya Yoshida (ed.), *Abscisic Acid: Methods and Protocols*,
Methods in Molecular Biology, vol. 2462, https://doi.org/10.1007/978-1-0716-2156-1,

MIX
Papier aus verantwortungsvollen Quellen
Paper from responsible sources
FSC® C105338

If you have any concerns about our products,
you can contact us on
ProductSafety@springernature.com

In case Publisher is established outside the EU,
the EU authorized representative is:
Springer Nature Customer Service Center GmbH
Europaplatz 3, 69115 Heidelberg, Germany

Printed by Libri Plureos GmbH
in Hamburg, Germany